Conformations

Conformations

Connecting the Chemical Structures and Material Behaviors of Polymers

Alan Tonelli
Jialong Shen

CRC Press is an imprint of the
Taylor & Francis Group, an **informa** business

CRC Press
Taylor & Francis Group
6000 Broken Sound Parkway NW, Suite 300
Boca Raton, FL 33487-2742

First issued in paperback 2021

ISBN 13: 978-1-138-57032-0 (hbk)
ISBN 13: 978-1-03-224153-1 (pbk)

DOI: 10.1201/b22496

Library of Congress Cataloging-in-Publication Data

Names: Tonelli, Alan E., author. | Shen, Jialong, 1987- author.
Title: Conformations : connecting the chemical structures and material behaviors of polymers / by Alan Tonelli and Jialong Shen.
Description: Boca Raton : CRC Press, [2020] | Includes bibliographical references and index.
Identifiers: LCCN 2019050313 (print) | LCCN 2019050314 (ebook) | ISBN 9781138570320 (hardback) | ISBN 9780203703601 (ebook)
Subjects: LCSH: Polymers.
Classification: LCC QD381 .T648 2020 (print) | LCC QD381 (ebook) | DDC 547/.7045--dc23
LC record available at https://lccn.loc.gov/2019050313
LC ebook record available at https://lccn.loc.gov/2019050314

Visit the Taylor & Francis Web site at
http://www.taylorandfrancis.com

and the CRC Press Web site at
http://www.crcpress.com

Publisher's Note
The publisher has gone to great lengths to ensure the quality of this reprint but points out that some imperfections in the original copies may be apparent.

Contents

Preface

Among the materials found in Nature's many diverse living organisms or produced by human industry, those made from polymers are dominant. In Nature, they are not only dominant, but they are, as well, uniquely necessary to life. Without proteins to give form and function to a myriad of living animals and plants, without DNAs and RNAs to direct the syntheses of their unique sets of necessary proteins, and without the polysaccharides (cellulose, amylose, chitin, etc.) necessary to all living species, life as we experience on earth could not exist. Additionally, the ever increasing and unabated differential margin of production enjoyed by industrial polymers over metals and ceramics continues.

The world dominance of polymer materials is a result of their structural features. Their high molecular weight, long, 1-dimensional, and flexible chains, which lead to highly variable sizes and shapes, produce behaviors and properties that are unique to them and their resultant materials. Because in almost all instances, at least some of the bonds constituting polymer backbones can be rotated into several distinct conformations, the overall collections of potential polymer chain conformations, sizes, and shapes are nearly incomprehensibly large. More important is the ability of polymers to select among their conformations, and adopt those able to provide an appropriate internal response to their external environments and the forces they experience. In other words, polymer chains can be thought to have an ***Inside*** "mind of their own", their conformational preferences, which determine the ***Outside*** responses of materials made from them.

Clearly, materials made from small molecules, atoms, or ions, do not have this internal conformational degree of freedom. This is the reason why polymers, with their long flexible chains, and their materials show a number of behaviors and properties that are unique to them. We call the unique properties displayed by nearly all polymers ***Polymer Physics*** and describe several of them in Chapter 1—"Polymer Physics or Why Polymers and Their Materials Can Behave in Unique Ways". The only means for materials made from small molecules, atoms, or ions to respond to their environments and/or the forces placed upon them is through alterations in their collective arrangements or overall configurations. To establish which of these molecular configurations are most favorable and therefore likely, requires solving a many-body problem, which for macroscopic material samples is not presently possible.

On the other hand, 50 years ago, Flory (Flory, P. J. [1969], *Statistical Mechanics of Chain Molecules*, Wiley-Interscience, New York) showed how the distinct chemical structures of polymers can be rigorously accounted for when establishing the conformational preferences of their chains. Adopting the rotational isomeric state (RIS) model of polymer conformations (Volkenstein, M. V. [1963], *Configurational Statistics of Polymer Chains*, translated by Timasheff, S. N. and Timasheff, N. J. from the 1964 Russian Ed., Wiley-Interscience, New York; Birshstein, T. M. and Ptitsyn, O. B. [1964], *Conformations of Macromolecules*, translated by Timasheff, S. N. and Timasheff, N. J. from the 1964 Russian Ed., Wiley-Interscience, New York), Flory pointed out that the likelihood or populations of local polymer chain conformations are generally only pair-wise nearest neighbor dependent. In other words, the energy of a particular backbone bond conformation depends only upon the conformations of the bond in question and those of its immediate neighbors. Consequently, the energy of each overall conformation of a polymer chain of $\boldsymbol{n}$ backbone bonds, $\boldsymbol{E}_{conf}$, is a summation of pair-wise-dependent local bond conformational energies, $\boldsymbol{E}_{conf} = \sum_{i=2}^{n-1} \boldsymbol{E}(\varphi_{i-1},\varphi_i)$, where $\boldsymbol{E}(\varphi_{i-1},\varphi_i)$ is the local conformational energy when bonds $\boldsymbol{i-1}$ and $\boldsymbol{i}$ adopt the conformations φ_{i-1} and φ_i.

If each polymer chain conformation $(\varphi_2,\varphi_3,\varphi_4,\ldots\varphi_{n-3},\varphi_{n-2},\varphi_{n-1})$ is considered a system, then the collection of all polymer chain conformations, $\boldsymbol{N}_{conf}$, may be considered an ensemble of statistical mechanical systems, each with an energy $\boldsymbol{E}_{conf}$. This leads to a soluble conformational partition function, $\boldsymbol{Z}_{conf}$, for the individual polymer chain and all its attendant thermodynamic properties, because their pair-wise-dependent system energies, $\boldsymbol{E}_{conf}$, are readily estimable (Hill, T. L. [1960], *An Introduction to Statistical Mechanics*, Addison-Wesley, Reading Massachusetts). In addition to their pair-wise-dependent conformational energies, because linear polymer chains are 1-dimensional systems, matrix multiplication methods were also developed by Flory (1969) for evaluating local and global chain properties, such as bond conformational populations and overall sizes [mean-square end-to-end distances ($<r^2>_o$) and radii of gyration ($<R_g^2>_o$)], with each appropriately averaged over all chain conformations $\boldsymbol{N}_{conf}$.

Of course $\boldsymbol{E}_{conf}$ depends on the detailed microstructure of each chemically different polymer, and so too do their flexibilities ($\boldsymbol{Z}_{conf}$), sizes, and shapes. As a consequence, the difference in behaviors and properties of materials made from chemically distinct polymers, which we call ***Polymer Chemistry***, are directly connected to the differences in their conformational preferences. In Chapter 2—"*Polymer Chemistry* or the Detailed Microstructures of Polymers", we establish the possible local polymer microstructures and later describe and demonstrate how they can be experimentally established.

As indicated below, we take advantage of this feature of polymer chains to connect the behaviors and properties of their materials to their detailed chain microstructures:

Microstructures → Conformations → Overall Sizes, Shapes & Flexibility → Behaviors & Properties

The key step or feature of this approach is establishing the conformational preferences of individual polymer chains as a function of their detailed chemical microstructures, as described at length in Chapter 3—"Determining the Microstructural Dependent Conformational Preferences of Polymer Chains." In Chapter 4—"Experimental Determination of Polymer Microstructures with ^{13}C-NMR Spectroscopy"—we show how the conformational characteristics of polymers are related to and used to assign their ^{13}C-Nuclear Magnetic Resonance (^{13}C-NMR) spectra, thereby permitting their microstructures to be determined by this experimental technique.

Finally in Chapter 5—"Connecting the Behaviors/Properties of Polymer Solutions and Liquids to the Microstructural Dependent Conformational Preferences of Their Polymer Chains"—and in Chapter 6—"Connecting the Behaviors/Properties of Polymer Solids to the Microstructural Dependent Conformational Preferences of Their Individual Polymer Chains"—we show which and how certain behaviors and properties of polymers and their materials can be connected to and explained in terms of their single chain microstructural-dependent conformations. In addition, we point out behaviors and properties dominated by interactions between polymer chains and between polymers and solvents, and, as a result, are not amenable to our ***Inside*** polymer chain microstructure ←→ ***Outside*** polymer material property approach. These include polymer rheology, process-dependent structures and behaviors, two-phase crystalline and amorphous morphologies, softening or glass-transition temperatures, in addition to others.

We close our discussion in Chapter 7—"Biopolymer Structures and Behaviors, with Comparisons to Synthetic Polymers"—by comparing and contrasting the behaviors of synthetic polymers, with their comparatively simple microstructures, to those of biopolymers, whose microstructures and functions are infinitely more diverse.

We cannot fail to mention that no previously published polymer science, chemistry, physics, engineering, or materials text, aside from *Polymers from the inside Out: An Introduction to Macromolecules* (Tonelli and Srinivasarao, 2001), takes realistic account of the microstructure-dependent conformational preferences of individual structurally distinct polymer chains when attempting to rationalize/understand the behaviors and properties of their materials. This despite the availability of conformational RIS models that have been or can readily be developed for most commercially important synthetic man-made polymers, and that are also relatively easy to implement.

The reality is that ***Polymer Chemistry*** is seldom considered when trying to understand the behaviors of polymer materials. This is made abundantly clear in a recent perspective titled "Polymer Conformation-A Pedagogical Review" published in *Macromolecules* (2017), doi:10.1021/acs.macromol.7b01518 and written by Professor Zhen-Gang Wang in celebration of the 50th anniversary of the journal *Macromolecules*. Despite its title, this paper is really a perspective/review of ***Polymer Physics***, because all polymers are considered to be randomly walking, freely rotating, or worm-like chains, without distinctive chemical microstructures that result in varying conformational preferences.

We fervently hope that, with the publication of our book, the neglect of polymer chemistry will no longer remain the case. Because without accounting for the conformational preferences of synthetic polymers, relevant relations between their distinct microstructures, behaviors, and material properties cannot be established, i.e., the relevance of the chemistry of individual polymer chains to the behaviors of their materials would remain largely a mystery.

To quote Flory, "Quantities such as molecular weight, radius of gyration, degree of branching, *etc.* that characterize the molecule as a whole do not suffice for an understanding of the virtually endless variations in the properties that distinguish one polymer from another. For this purpose it is necessary to examine the chemical structures of the constituent units, taking into account geometric parameters pertaining to chemical bonds, the conformations accessible to the chain skeleton and the resulting spatial configurations" (Flory, 1969).

It is more than obvious that our approach to the behavior of polymer materials, whose polymer chains can be thought to have an ***Inside*** "mind of their own", their conformations, which determine the ***Outside*** responses of materials made from them, could not have been taken without the knowledge in Flory's *Statistical Mechanics of Chain Molecules*. So on the golden anniversary of its publication, we gratefully express our gratitude for its many lessons.

Authors

Alan Tonelli, born in Chicago in 1942, received a BS in Chemical Engineering from the University of Kansas, in 1964 and a PhD in Polymer Chemistry from Stanford in 1968, where he was associated with the late "Father of Polymer Science" and Nobelist Professor Paul J. Flory. He was a member of the Polymer Chemistry Research Department at AT&T-BELL Laboratories, Murray Hill, NJ for 23 years. In 1991, he joined the Textile Engineering, Chemistry, & Science Department and the Fiber & Polymer Science Program in the College of Textiles at North Carolina State University in Raleigh, where he is currently the INVISTA Prof. of Fiber & Polymer Chemistry. Professor Tonelli's research interests include the conformations, configurations, and structures of synthetic and biological polymers, their determination by NMR, and establishing their effects on the physical properties of polymer materials. More recently, the formation, study, and use of inclusion complexes formed with polymers and small molecule guests, with hosts such as urea and cyclodextrins, to nanostructure and safely deliver biologically active molecules to polymer materials have been the focus of his research.

Jialong Shen, born in Hangzhou, China, in 1987, received a PhD in Fiber and Polymer Science from North Carolina State University, North Carolina, United States, in 2017. His research interests include the molecular basis of polymer glass transitions, host-guest supramolecular chemistry, and the applications of bio-macromolecules such as carbohydrate polymers and enzymes. He is currently a postdoctoral research scholar in the Textile Engineering, Chemistry, & Science Department at North Carolina State University.

1 *Polymer Physics* or Why Polymers and Their Materials Can Behave in Unique Ways

INTRODUCTION

In this introductory chapter, we seek to describe polymers in only a minimal generic way that distinguishes them from small molecules, i.e., as long, high molecular weight, and flexible chains. Our purpose here is to demonstrate why these unique structural features alone result in behaviors and properties that are unique to polymers and the materials made from them. We call this *Polymer Physics*, because these unique polymer behaviors are manifested by virtually all polymers regardless of their detailed chemical microstructures. Of course the degree to which polymers with distinct microstructures evidence their unique *Polymer Physics* behaviors does depend on their detailed chemical microstructures, which we call *Polymer Chemistry* and is the main emphases of the following chapters.

If we compare the size of an ethane molecule $CH_3–CH_3$ with a molecular weight (MW) = 30 g/mol to a polyethylene chain [PE = $(-CH_2–CH_2-)_n$-], with n = 25,000 repeat units and a MW = 700,000 g/mol, which is not atypical, after magnifying each of them a billion times, ethane would be about a foot long and the fully extended all *trans* PE chain more than 4 miles in length. Clearly, polymers are in fact large molecules or **macromolecules**, and the following demonstration clearly shows how the difference in sizes between typical small molecules and long high molecular weight polymer chains can dramatically affect their behaviors.

We use the Cannon-Ubbelohde viscometer shown in Figure 1.1 to measure the flow times of three liquids: water, a 0.5 wt% aqueous solution of ethanol E = [$CH_3–CH_2–OH$], and a 0.5 wt% aqueous solution of poly(ethylene oxide) [PEO = $–(CH_2–CH_2\text{-}O\text{-})_n$-], with a molecular weight of 4,000,000 g/mole and containing n = 90,909 repeat units, through the viscometer's capillary. You'll note that the repeat unit of PEO is closely similar to that of E. This choice was made to alleviate, as much as possible, differences in their chemistry, so that they are distinguished solely by their sizes.

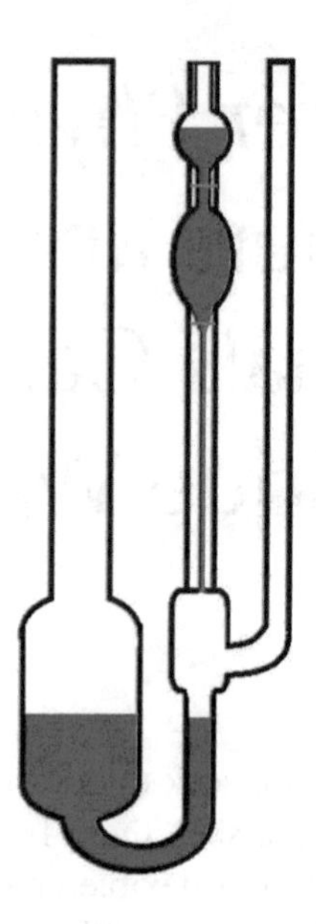

FIGURE 1.1 Cannon-Ubbelohde viscometer filled with a liquid. (Reprinted with permission from Shen, J. and Tonelli, A. E., *J. Chem. Educ.*, 94, 1738–1745, 2017. Copyright 2017 American Chemical Society.)

Flow times observed in the #1 Cannon-Ubbelohde viscometer, shown in Figure 1.1, were, respectively, ~108, 107, and 343 seconds for water, 0.5% ethanol in water, and 0.5% PEO (M = 4,000,000) in water. (It is also possible to use a less expensive Fenske viscometer. See a video demonstration of us using a Cannon-Ubbelohde viscometer in the supporting information of Shen and Tonelli 2017.)

Division of the MW of PEO by the MW of E reveals the ratio of the number of small E molecules to the number of long PEO chains, i.e., 4,000,000/46 = 87,000. There are 87,000 times more ethanol molecules than PEO chains in their 0.5 wt% aqueous solutions. Clearly, the much larger randomly coiling PEO chains dramatically slow the flow of the water molecules even though they constitute only 0.5 wt% of their aqueous solution, while the same weight and much greater number of small E molecules has virtually no effect on the flow of its 0.5 wt% aqueous solution. Later, in Chapter 5, we learn that even at only 0.5 wt%, the repeat units of the large randomly coiling PEO chains are in close contact with all of the 99.5 wt% solvent water molecules, thereby retarding their flow through the viscometer's narrow capillary.

Our 2nd demonstration of *Polymer Physics* uses a simple rubber band to illustrate some unique behaviors of a network formed by cross-linking mobile liquid polymer chains. To form an elastic network that can easily be stretched and then reversibly returned to its unstretched state, i.e., to its unstretched size and shape, when the stretching force is removed, three requirements must be met: (i) polymer chains, which can alter their sizes and shapes; (ii) the temperature must be higher than the softening temperatures (T_g and/or T_m) of the amorphous and/or crystalline regions of the polymer sample, because polymer chains must be mobile and able to change their conformations when the network is stretched; and (iii) to eliminate irreversible movements or flow of the polymer chains, they must be cross-linked into a network.

The volume of a stretched polymer network (see Figure 1.2b) differs little from the volume of the unstretched network (see Figure 1.2a). This can be confirmed visually by stretching a thick rubber band and noting that while its length obviously increases on stretching, both its width and thickness decrease. Consequently, the at-rest and stretched rubber bands have closely similar volumes. This implies that the average separation between chains remains constant, and therefore the interactions between polymer chains are not altered, suggesting that the change in interchain energy ΔU (interchain) ~ 0 upon network stretching.

In addition, as a comparison of Figures 1.2a and b makes plain, some of the chains between cross-links are extended, while some are compressed when the rubber network is uniaxially stretched. Stretching a polymer network clearly must cause the conformations of the chains between cross-links to change (See Figure 1.3 for examples of PE chain conformations, both extended and compact). In Chapter 3, we discuss the conformations of polymer chains in some detail.).

In Chapters 3 and 6, we will illustrate that large changes in the distances between cross-links can be achieved by bond rotations about a small number

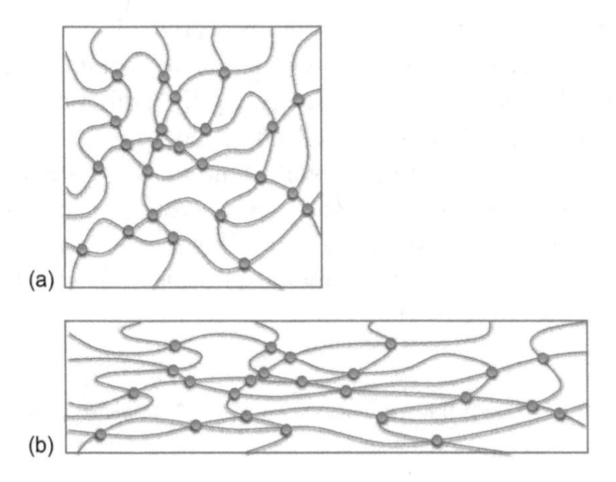

FIGURE 1.2 Cross-linked network (a) in unstrained at rest and (b) in strained states. (Reprinted with permission from Shen, J. and Tonelli, A. E., *J. Chem. Educ.*, 94, 1738–1745, 2017. Copyright 2017 American Chemical Society.)

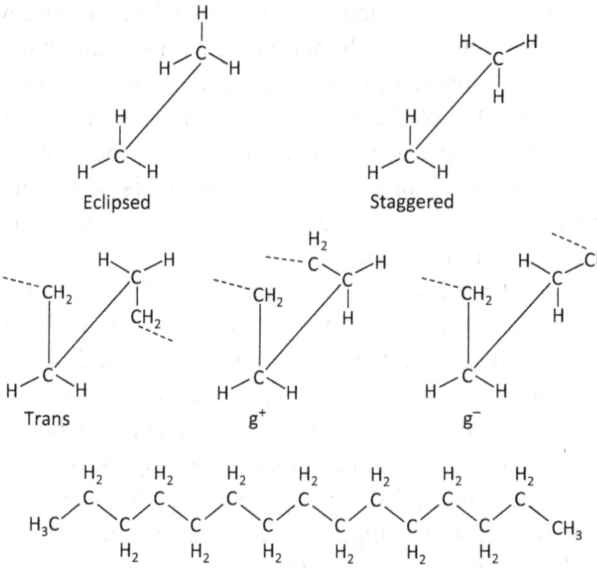

FIGURE 1.3 Sawhorse projections of ethane conformers (top), of trans and gauche conformers in PE (middle), and PE in the extended all trans conformation (bottom).

of the backbone bonds between cross-links. Because the number of bonds changing their conformations as the network is stretched is small in comparison to the total number of backbone bonds between cross-links, the change in polymer chain conformational energy is small or ΔU (intrachain) ~ 0.

Since both inter- and intrachain contributions to the total change in internal energy that accompany network stretching are ~0, we can assume network stretching is characterized by an overall ΔU ~ 0. The first law of thermodynamics states that the energy of a system U can only be changed, raised or lowered, respectively, by the addition or removal of heat (Q) or by work (W) performed on or done by the system. Thus, ΔU = Q + W, with heat into and work done on the system, raising the system's energy. Stretching of a polymer network is accompanied by a ΔU ~ 0, so Q must equal –W or –Q = W. This means that the work done on a polymer network upon stretching must be released in the form of heat to maintain an overall ΔU ~ 0.

```
                                                    ~~(CH2   CH2)n ~~
                                                         \   /
Placing a large (>#64) natural rubber band (cis-1,4-polyisoprene =C=C )
                                                         /   \
                                                       CH3    H
```

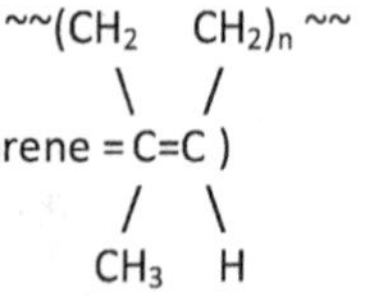

between your wet lips and quickly extending it will cause the heat escaping from the rubber band upon stretching to be detected by a warming of your lips. (When doing this wear safety glasses and use sterile rubber bands, which are available.) The stretched rubber band may alternatively be placed between your wet lips and allowed to rapidly contract. In this instance, your wet lips should be able to sense the cooling produced by the heat absorbed from them by the contracting rubber band, counteracting the loss of energy produced by the work performed by the contracting rubber band. (See how we do it in the video included in the supporting information of Shen and Tonelli 2017.).

We have strongly suggested based on our rubber band demonstration that a stretched elastic polymer network does not contract and return to its at-rest unstretched state in order to lower its energy, because upon stretching or contraction $\Delta U \sim 0$. So what must be the origin of the restoring force? If it is not lowering the energy of the rubber, it must be increasing the entropy accompanying network contraction. Since stretching and contraction of the elastic polymer network does not result in a change in volume or in network density, this cannot be the source of the change in network entropy upon extension or contraction. The only remaining source for the entropy change which causes the stretched network to contract when the extensional force is removed must be associated with changes in conformations of the polymer chains between cross-links upon extension and retraction. Whether the distances between network crosslinks are increased or decreased upon extension of the network, the populations of the favorable conformations available to the stretched or compressed network chains are reduced. This reduction in conformational entropy upon extension is reversibly increased and recovered upon contraction as the chains resume sampling their complete equilibrium populations of conformations.

Our final demonstration of behaviors unique to polymers or *Polymer Physics* employs a 5% solution of poly(vinyl alcohol) [PVOH = $-(-CH_2-\underset{\underset{OH}{|}}{CH}-)_n-$] in water, to which a small amount of a 5% aqueous solution of borax (sodium borate = $Na_2B_4O_7 \times 10H_2O$) has been added dropwise with stirring to form a gel often called "slime." (Neither PVOH nor borax are toxic through handling, but wash hands thoroughly to avoid any chance of ingestion. See Casassa et al. 1986.)

Figure 1.4, shows the borate ions [$B(OH)_4^-$] generated upon dissolution of the borax are capable of interacting with the hydroxyl groups on the PVOH chains. As a result, non-covalent tetra-functional cross-links are formed, but are only weak physical attractions, which may be easily disrupted if the gel is strained.

The time-dependent behaviors/responses of materials made from long-chain flexible polymers are a unique and characteristic aspect of ***Polymer Physics***. These properties can be easily demonstrated through the following

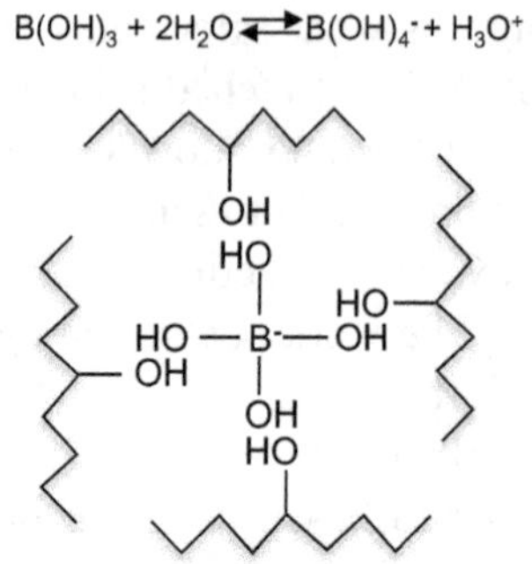

$$B(OH)_3 + 2H_2O \rightleftharpoons B(OH)_4^- + H_3O^+$$

FIGURE 1.4 Schematic diagram showing the ability of borate ions produced in aqueous solutions of borax to form temporary, physical cross-links in PVOH solutions resulting in dilute "slime" gels.

three simple slime demonstrations. (See demonstration video included in the supporting information of Shen and Tonelli 2017 for explanations of the slime responses to forces applied for varying times.)

When the slime gel is removed from the beaker, rolled into a cylinder, and a portion is hung over the edge of a desktop, a single continuous strand forms and flows until it reaches the floor and puddles up. The relative weakness of the interactions between the borate ions and the PVOH hydroxyls are evidenced by this behavior. The single flowing continuous strand of slime also illustrates that, though dynamic, an overall cross-linked network is maintained due to the weak and temporary nature of the cross-links in slime. These cross-links, though not able to withstand the force of gravity, are continuously broken and reformed, allowing irreversible flow of the PVOH chains passed each other.

Next, we roll the slime gel into a ball and, from a height of several feet, drop it on a desktop. The slime ball bounces elastically several times before sticking to, spreading on, and forming a puddle on the desktop.

In our final slime observation, we once again roll the slime into a cylinder and pull it rapidly on both ends in an attempt to extend it very abruptly. The slime cylinder fails and breaks in a brittle manner if pulled quickly enough, with very little extension and sharp, flat fracture surfaces reminiscent of a broken icicle. A similar demonstration of brittle slime behavior induced by the application of large high frequency stresses can be accomplished by throwing a ball of slime very hard against a solid wall. This results in its fracture into a large number of small pieces, which are only recovered with tedious effort, so this demonstration should only be conducted on an outside wall.

In the three "experiments," the slime gel responded in very distinct ways when stressed differently. A cylinder of slime hung over the edge of a desktop flowed irreversibly as it was subjected to the constant application of gravity. The slime ball dropped onto the desktop bounced elastically several times before puddling up. The slime cylinder that was rapidly extended was observed to fail in a brittle manner. To summarize and conclude, the manner in which the slime gel responded was, in addition to its magnitude, sensitively dependent on the frequency of the applied force, i.e., the amount of time the force was applied.

Each of the behaviors briefly demonstrated in these ***Polymer Physics*** demonstrations are unique to polymers and their materials. Detailed reasons for each are subsequently provided in Chapters 5 and 6, where the magnitudes of their responses are seen to be dependent on their chemical microstructures and the conformational preferences resulting from them. There, we begin to rationally understand the ***Inside*** "minds of their own," i.e., their microstructural dependent conformational preferences, and how these are connected to and determine the ***Outside*** responses of polymer materials.

REFERENCES

Casassa, Z., Sarquis, A. M., Van Dyke, C. H. (1986), *J. Chem. Educ.*, 63 (1), 57.
Shen, J., Tonelli, A. E. (2017), *J. Chem. Educ.*, 94 (11), 1738–1745.

DISCUSSION QUESTIONS

1. What is the principal reason that polymers and their materials show behaviors distinct from materials made from small molecules and atoms?
2. How can the addition of such small amounts of polymer solutes (<1 wt%) to small molecule solvents so dramatically slow down the flow of their solutions?
3. Given no further information, how would you expect polypropylene with a MW = 42,000 and poly(vinyl chloride) with a MW = 62,000 to affect the viscosities/flow times of their solutions?
4. Why did we conclude that the resistance to stretching a rubber band is entropic in origin?
5. As we stretch a cross-linked polymer network, such as a rubber band, some chains between cross-links are stretched, while others are shortened or compressed. How does this effect the number of conformations they can adopt, i.e., their conformational entropy?
6. What do the rubber-band and slime DEMOS have in common, and how do they differ?
7. We observed that even though the slime solution and gel that was formed upon addition of a small amount of borax contained only ~5% PVOH, its responses to deformation were very dependent on the frequency or time of the applied force. Discuss the reason(s) why?

2 *Polymer Chemistry* or the Detailed Microstructures of Polymers

POLYMERIZATION

Step-Growth Polymers

Polymers are made by linking together small molecules called monomers. All monomers must possess the ability to undergo at least two linking reactions to be inserted into and eventually produce a polymer chain. Linear polymers result from the exclusive linking together of bi-functional monomers. As examples, single molecules containing two functional groups -X and Y- that react with each other to form a covalent link -Z-, or two different molecules each with a pair of reactive function groups (X-R-X) and (Y-R′-Y) that also form covalent links -Z- can be used to make polymers produced in a Step-Growth fashion:

$$\mathbf{n}(\mathbf{X\text{-}R\text{-}Y}) \rightarrow \mathbf{X\text{-}(\text{-}R\text{-}Z\text{-})_{n-1}\text{-}Y} \text{ or } \mathbf{n}(\mathbf{X\text{-}R\text{-}X}) + \mathbf{n}(\mathbf{Y\text{-}R'\text{-}Y}) \rightarrow$$

$$\mathbf{X\text{-}(\text{-}R\text{-}Z\text{-}R\ \text{-}Z\text{-})_{n\text{-}1}\text{-}Y}$$

For instance, the common classes of synthetic polymers, polyesters, and polyamides, are made from monomers where X and Y are a carboxylic acid group and a hydroxyl or an amine group, respectively. Proteins are synthesized from bifunctional α,ω-amino acids ($H_2N\text{-}CH(R)\text{-}C(=O)\text{-}OH$), with 20 different R groups (see Chapter 7).

The polyesters and polyamides are generally termed Step-Growth polymers because of the mechanism or way they grow from a large collection of monomers into a sample with many fewer, but much longer polymer chains. A collection of polymers with a reasonably high average molecular weight (MW) is not achieved until nearly all of the functional groups (-X, Y-) have been converted to -Z- links (Carothers 1936; Flory 1953). The fact that the reactivity of, or reaction rate between, -X and Y- functional groups is generally independent of the sizes of the molecules they are attached to means that the probability of forming oligomers or chains of varying lengths depends only on the concentrations of the two molecular reactants that come together to form the product molecule (Flory 1953; Odian 2004).

Of course until nearly all functional groups have reacted, the concentrations of relatively short reactants are high, while at the same time the concentrations of the longer reactants are low. It is only when nearly all -X and Y- reactants have come together to form -Z- links, and the concentrations of longer reactants are now reasonably high, does the average MW or number of repeat units, $\overline{X}_n$, become significant, i.e., $\overline{X}_n = 1/(1 - p)$, where p is the fractional extent of functional groups reacted (Carothers 1936; Flory 1953). Hence, the name Step-Growth polymerization to characterize the step-by-step progression from a collection of many monomers to a sample of many fewer polymer chains with a reasonably high average MW. An example of this progression is schematically presented in Figure 2.1.

Because of their mode of polymerization (-X + Y- → -Z-), the chemical microstructures of Step-Growth polymers are generally not variable, but rather are easily and reliably predicted. Of course complicated Step-Growth polymers can be purposefully made by using several monomers having the same functional groups, but attached to a variety of non-reactive fragments. For example, if all of the Step-Growth monomers X-R-X, X-R′-X, X-R″-X, Y-R-Y, Y-R*-Y, and Y-R**-Y were allowed to simultaneously react, a Step-Growth copolymer with a variety of different "-Z- links" (R-Z-R, R-Z-R*,

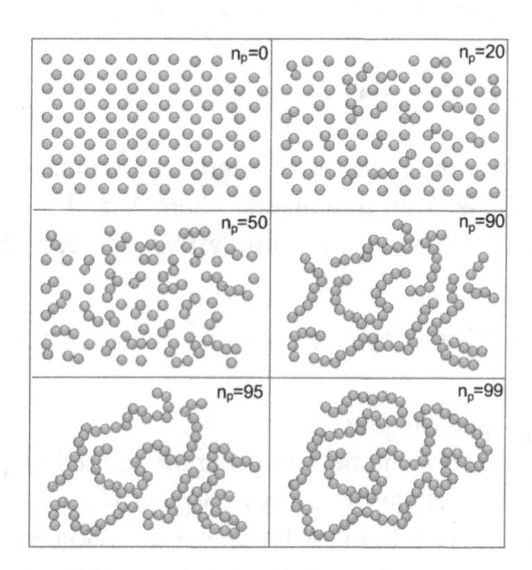

FIGURE 2.1 Microscopic snapshots at different extents or number of reactions (n_p) resulting from a trial reaction of 100 X-R-Y monomers. Note that single beads either represent monomers or repeat units. (Reproduced with permission from Shen, J. and Tonelli, A. E., *J. Chem. Ed.*, 94, 1738–1745, 2017. Copyright 2017 American Chemical Society.)

R-Z-R**, R′-Z-R, R′-Z-R*, R′-Z-R**, R″-Z-R, R″-Z-R*, R″-Z-R**) and complex sequences of these enchained monomers would result.

Chain-Growth Polymers

The other major method for forming polymers is called Chain-Growth polymerization, because once a chain is initiated and starts to grow, it grows rapidly and is terminated before a significant fraction of remaining monomers have reacted. Though initially a few chains with high MWs are rapidly grown, it takes considerably more time for all of the monomers to be initiated, form chains, and achieve a polymer sample with a high average MW.

The most common monomers used to make Chain-Growth polymers are unsaturated with at least one reactive double bond, as, for example, in vinyl monomers like $CH_2{=}CH(R)$. Note from the drawings of the simplest vinyl monomer ethene in Figure 2.2, the four electrons shared between carbon atoms form two distinctly different bonded pairs, the σ- and π-electrons. While the shared σ-electron pair is localized closely to and held strongly

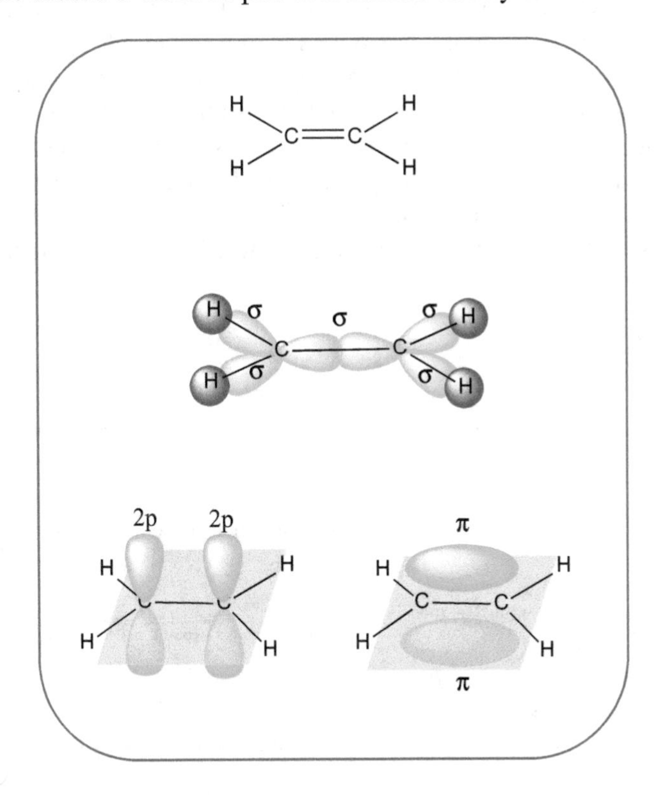

FIGURE 2.2 Ethene line drawing (top), σ-electron bonding (middle), and π-electron bonding (bottom).

between the pair of carbon atoms, the pair of π-electrons are by comparison located far above or below and loosely held by the carbon atoms sharing them. Consequently, the π-electrons are susceptible to attack by and reaction with a variety of active species, such as free-radicals and ions, and may be shared, incorporated into, and form two σ-bonds.

A reactive species R* (a free-radical or ion) can attack the π-bond of a vinyl monomer by opening it and sharing one of the π-electrons to form a new σ-bond and form a reactive species, $R^* + CH_2{=}\underset{|}{\overset{X}{CH}} \rightarrow R\text{-}CH_2\text{–}\underset{|}{\overset{X}{C^*}}$, which can then rapidly react with and add many monomers and grow in a Chain-Growth manner.

Chain-Growth Polymer Microstructures

Unlike Step-Growth polymerization, which depends on specific reactions between functional groups to similarly link all the monomers into polymer chains, vinyl monomers may be incorporated into growing polymer chains in a number of different ways. For instance, as shown below in Figure 2.3, vinyl monomers usually have a direction and may be enchained in two distinct ways. Except for monomers of the type $CH_2{=}CHR_2$ with two identical substituent side-chains, vinyl monomers with a single side-group substituent can also produce polymer chains with a variety of stereosequences or tacticities (see Figure 2.3). When coupled with the potential for monomer incorporation with different regiosequences, vinyl polymer chain tacticities can potentially lead to a large number of complicated local polymer microstructures, which need to be characterized before we can hope to establish structure-property relations for materials made from them.

```
        H   H
         \ /
growing chain ~~~•  +  C=C  →
         / \
        H   R,

 H     H           H     H
  \   /             \   /
~~~ C -- C•   or  ~~~ C -- C•
  /   \             /   \
 H     R           R     H
  (T-H)              (H-T)

        Regular

(T----H)  (T----H)  (T----H)  (T----H)
-CH2-CHR- CH2-CHR-CH2-CHR-CH2-CHR-

        Irregular

(T----H)  (T----H)  (H-----T)  (T-----H)
-CH2-CHR- CH2-CHR-CHR-CH2-CH2-CHR-
```

(a) Regiosequences: Directions of monomer insertion

```
~~~CH2-CH-CH2-CH-CH2-CH-~~~ → Isotactic (All m-diads)
       |  m  |  m  |
       R     R     R

             R
         r   |   r
~~~CH2-CH-CH2-CH-CH2-CH-~~~ → Syndiotactic (All r-diads)
       |           |
       R           R

             R
         r   |   r
~~~CH2-CH-CH2-CH-CH2-CH-CH2-CH~~ → Atactic or random
       |           |     |          r- and m-diads
       R           R  m  R

m = a meso diad (same attachment)
r = a racemic diad (opposite attachment)
```

(b) Stereosequences of attached side-chains

FIGURE 2.3 Vinyl polymer microstructures: (a) regiosequences and (b) stereosequences with regioregular T-H direction of monomer insertion.

When dienes are employed as monomers for Chain-Growth polymerization, there are a huge potential number of distinct ways they may be enchained to produce different microstructures, and these are indicated in Figure 2.4.

Example Monomer:

```
CH2=C–C=CH2
    |  |
    R  H
```

Let's consider the simpler symmetrical diene monomer 1,3-butadiene (CH_2=CH-CH=CH_2) and determine how many distinct microstructures it can produce during Chain-Growth polymerization (see Figure 2.5).

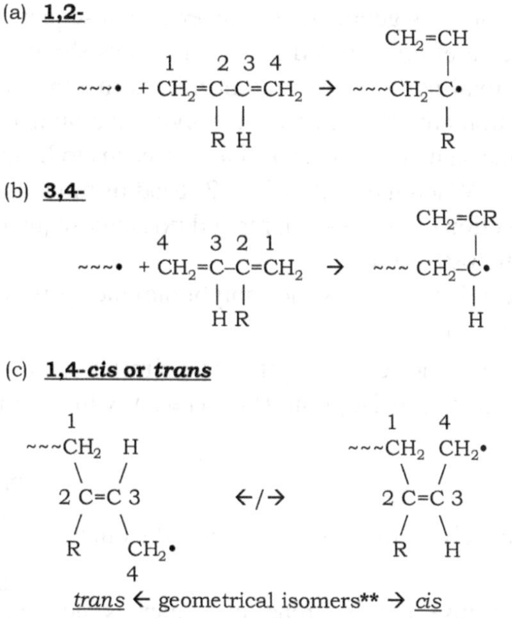

(a) **1,2-**

```
                                  CH2=CH
         1    2 3 4                  |
~~~• + CH2=C–C=CH2  →   ~~~CH2–C•
              |  |                   |
              R  H                   R
```

(b) **3,4-**

```
                                   CH2=CR
         4    3 2 1                   |
~~~• + CH2=C–C=CH2  →    ~~~ CH2–C•
              |  |                    |
              H  R                    H
```

(c) **1,4-*cis* or *trans***

```
     1                                   1     4
~~~CH2   H                        ~~~CH2  CH2•
     \  /                                \  /
    2 C=C 3          ←/→               2 C=C 3
     /  \                                /  \
    R    CH2•                           R    H
          4
```

trans ← geometrical isomers** → *cis*

FIGURE 2.4 (a–c) Potential diene monomer Chain-Growth additions. **IUPAC has decided to call *cis* and *trans* geometrical isomers *cis* and *trans* stereoisomers. However, we prefer and use the former terminology. (IUPAC, 1976.)

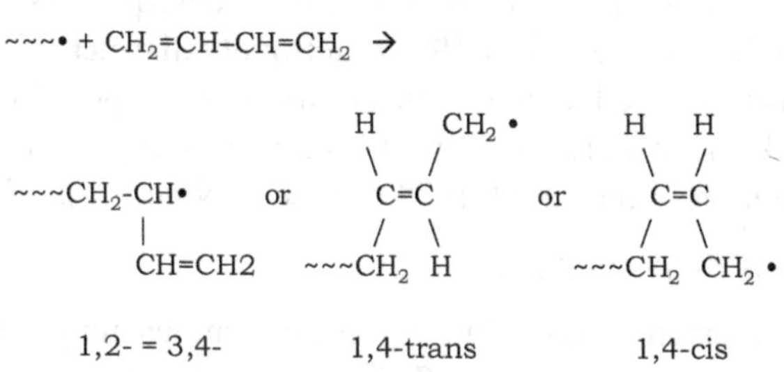

```
~~~• + CH2=CH–CH=CH2  →

                          H     CH2 •           H    H
                           \   /                 \  /
~~~CH2-CH•       or         C=C          or       C=C
       |                   /   \                 /  \
    CH=CH2          ~~~CH2   H            ~~~CH2   CH2 •

   1,2- = 3,4-           1,4-trans               1,4-cis
```

FIGURE 2.5 Potential 1,3-butadiene monomer Chain-Growth additions.

For example, how many distinct polybutadiene triads, i.e., three consecutive butadiene repeat units that are structurally different, can be produced?

Due to its symmetry, 1,2- and 3,4-additions are indistinguishable. Thus, each repeat unit can undergo 1,2- or 1,4-addition. For each 1,2-addition, two monomer insertion directions or regiosequences are possible. Stereosequence comes into play between each pair when there are two or more 1,2 additions. For the (1,2)(1,2)(1,2) triad, we have four possible stereosequences and four distinct out of eight total regiosequences (see Table 2.1). The possible monomer insertions for polybutadiene triads can be enumerated and are presented in the first column of Table 2.1, where six out of the eight drawn are distinct. For each of these monomer insertions, 1,4 *trans* and *cis* geometrical isomers, regiosequences, and stereosequences were carefully counted and are listed as shown on each row of Table 2.1. The total number of different microstructures for each distinct monomer insertion are obtained by multiplying the numbers of each type of isomers on the same row. One special case needs to be discounted from the total number. When the (1,2)(1,2)(1,2) triad of monomers are inserted in the same direction (⟶⟶⟶), mr and rm stereosequences do not produce distinct microstructures.

Thus, the total # of possible polybutadiene (PBD) triad microstructures = 61.

We leave it up to the reader to attempt a similar exercise for the more complex diene monomer isoprene (CH_2=C-CH=CH_2), which in natural rubber

```
      |
     CH3
```

is enchained entirely as the 1,4-*cis* structural isomer

```
~~~CH2   CH2~~~
     \   /
      C=C       .
     /   \
   CH3    H
```

Note that because the isoprene monomer is not symmetrical like 1,3-butadiene, it may be enchained in two distinct directions or regiosequences even with 1,4-addition. Also, when enchained as the 1,4-*cis* or *trans* geometrical isomers, all the atoms in the 1,4-repeat unit are confined to the same plane because of sp^2 bonding, and so the methyl side-chains cannot produce or be the source of different stereosequences.

It turns out that there are over **400 structurally distinct triads possible for polyisoprenes**. The fact that natural rubber is composed of all 1,4-*cis* units is remarkable and clearly points to Nature's ability to carefully direct the polymerization of the structurally precise polymers it produces.

Branching and Cross-Linking

Inadvertently created side-chains of various lengths may be formed in Chain-Growth polymer samples. This can happen by a chain-transfer mechanism that moves or transfers the unpaired electron or site of monomer

TABLE 2.1

Polybutadiene Triads Microstructures

Mode of Monomer Insertion	1,4 Addition	1,2 Addition			
List of possible triads	**Geometrical isomers[a]**	**Regiosequence[b]**	**Stereosequence[c]**	**Sum**	
(1,2)(1,2)(1,2)	None, so a 1 is used here	→→→=←←←, →←→=←→←, ←→→=→←←, →→←=←←→ 4	mm, mr, rm, rr, 4, 3	$3 \times 4 + 1 \times 3 =$	15
(1,2)(1,2)(1,4) = (1,4)(1,2)(1,2)	c, t 2	→→,→←, ←→ 3	m, r 2	$2 \times 3 \times 2 =$	12
(1,2)(1,4)(1,2)	c, t 2	→→, →←, ←→ 3	m, r 2	$2 \times 3 \times 2 =$	12
(1,2)(1,4)(1,4) = (1,4)(1,4)(1,2)	cc, ct, tc, tt 4	→,← 2	None, so a 1 is used here	$4 \times 2 \times 1 =$	8
(1,4)(1,2)(1,4)	cc, ct, tc, tt 4	→,← 2	1	$4 \times 2 \times 1 =$	8
(1,4)(1,4)(1,4)	ccc, cct = tcc, ttt, ttc = ctt, tct, ctc 6	1	1	$6 \times 1 \times 1 =$	6
Total					**61**

[a] *cis* (c) or *trans* (t).

[b] Insertion direction: tail to head (T⟶H) or head to tail (H⟵T).

[c] meso (m) or racemic (r) between each pair.

addition from the growing end of the same or another growing chain to an interior repeat unit. This is accompanied by abstraction of a proton from the new site of growth by the free-radical at the prior site of growth on the growing chain end (Odian 2004). When chain-transfer occurs from the growing end to the interior of the same chain, complex short chain branches can result as indicated in Figure 2.6.

Polymers may also be cross-linked after they are polymerized. Figure 2.7 presents examples of both branched and cross-linked polymers. The average

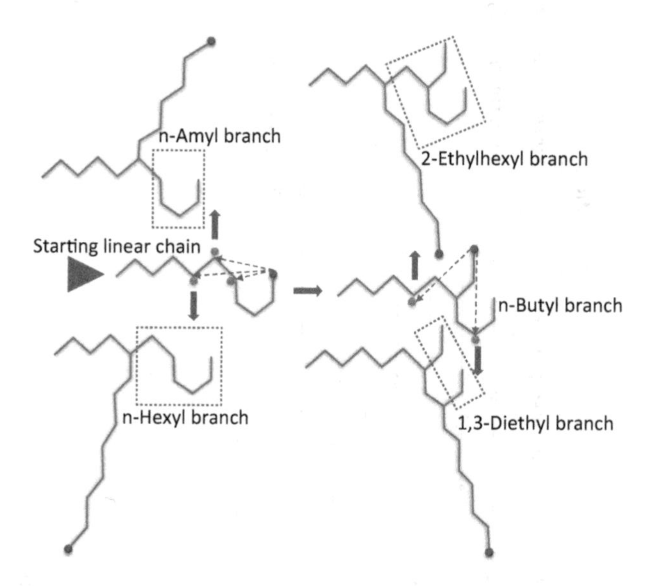

FIGURE 2.6 Intra-chain transfers producing a variety of short-chain branches. (Odian, G., *Principles of Polymerization*, 4th ed., Wiley, New York, 2004.)

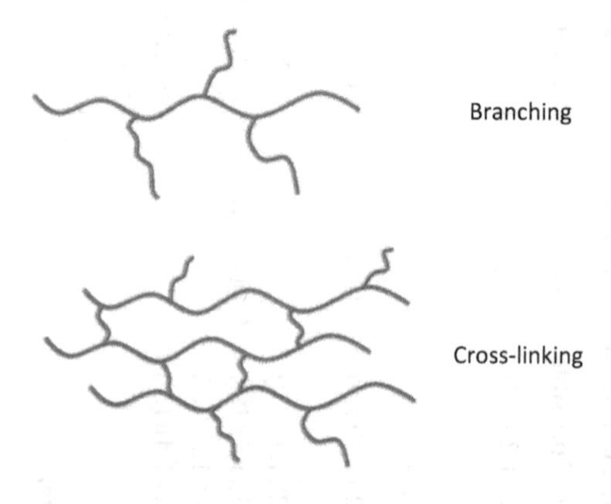

FIGURE 2.7 Examples of branched and cross-linked polymers.

numbers of branches or cross-links may be determined, and if branches are sufficiently short, they may be characterized. On the other hand, it is very difficult to determine the spacing between branches or crosslinks, or for that matter, the lengths of long branches, as well as their distributions.

As should now be apparent, Chain-Growth polymerizations of unsaturated monomers can produce a large variety of distinct chemical microstructures even if only a single vinyl or diene monomer is used. And as we will demonstrate in the following chapter, each distinct microstructure produces a local conformational consequence and alters the overall sizes, shapes, and flexibilities of polymer chains. The material behaviors are likewise similarly dependent on the chemical microstructures of their constituent polymer chains, including their types, their amounts, and their locations along the polymer's backbone.

Comonomer Sequences

If more than a single type of unsaturated monomer (vinyl or diene) is used in a Chain-Growth polymerization, then a co-, ter-, tetra-, etc. polymer containing 2, 3, 4, and more chemically different repeat units will be produced. As in polymers found in nature, specifically in proteins and DNAs and RNAs, potentially of greater importance to the behaviors of synthetic polymers than the amount of each monomer contained, is the sequence of monomers, i.e., how they are enchained with respect to each other. Figure 2.8 shows, schematically, copolymers with different co-monomer sequences.

In the next chapter, we address the important issue of determining the conformational characteristics of polymer chains with specified microstructures. This is the information that is pivotal in determining the connection between the ***Inside*** structures of polymers and the ***Outside*** behaviors/properties of their materials (Tonelli 2001), as well as their microstructures (Tonelli 1989).

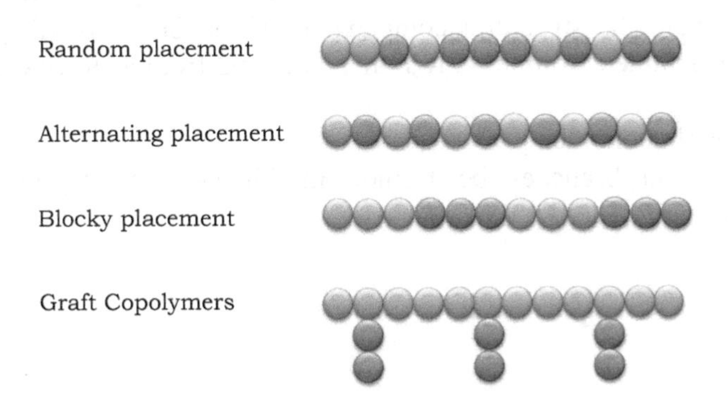

FIGURE 2.8 Copolymer monomer sequences.

REFERENCES

Carothers, W. H. (1936), *Trans. Faraday Soc.*, 32, 39–53.
Flory, P. J. (1953), *Principles of Polymer Science*, Cornell University Press, Ithaca, NY.
Odian, G. (2004), *Principles of Polymerization*, 4th ed., Wiley, New York.
Shen, J., Tonelli, A. E. (2017), *J. Chem. Ed.*, 94, 1738–1745.
Tonelli, A. E. (1989), NMR Spectroscopy and Polymer Microstructure: The Conformational Connection, Wiley, New York.
Tonelli, A. E. (2001), *Polymers from the Inside Out: An Introduction to Macromolecules*, Wiley-VCH, New York.

DISCUSSION QUESTIONS

1. Why do we call the polyester made from ethylene glycol (HO-CH_2-CH_2-OH) and adipic acid (HO-C(=O)-CH_2-CH_2-C(=O)-OH) (polyethylene adipate) a Step-Growth polymer?
2. If we start with N_0 X-R-Y monomers, how does the expected number average degree of polymerization, $\overline{X}_n$, vary with the fraction, p, of X- and -Y functional groups that have reacted. $\overline{X}_n$ must be given by the number of molecules initially present divided by the number of molecules remaining after a fraction p have reacted.
3. Unlike Step-Growth monomers that react on contact when molten or dissolved, vinyl monomers (CH_2=CH(R)) typically used for obtaining polymers *via* Chain-Growth polymerization must first be activated before adding additional monomers. In addition to differences in monomer reactivity, please compare and contrast the time dependences of monomer exhaustion, production of high molecular weight polymer chains, and achievement of a high average sample molecular weight for Step- and Chain-Growth polymerizations.
4. Compare the structural complexities of the Step-Growth polymer obtained from an X-R-Y monomer and from a CH_2=CH(R) Chain-Growth monomer.
5. How can branches be formed in Chain- and Step-Growth polymers?

3 Determining the Microstructural Dependent Conformational Preferences of Polymer Chains

INTRODUCTION

Virtually all polymers possess at least some backbone bonds about which rotation can take place between several stable low-energy conformations to alter the geometrical relationships between their constituent atoms. Bond rotation

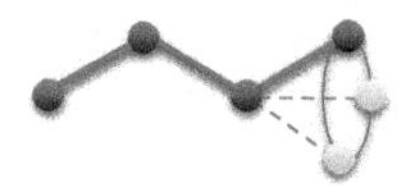

requires much less energy than bond stretching or valence angle bending, and the overall sizes and shapes of polymers are affected much more by bond rotations (see Figure 3.1). As a result, we may usually be able to specify the conformation of a polymer exclusively through determination of the rotation angle adopted by each of its backbone bonds, while maintaining constant bond lengths and bond valence angles. It is the ability of polymers to respond conformationally to their environments and the forces they experience, which gives them a unique and important intramolecular or *INSIDE* means to evidence unique *OUTSIDE Polymer Physics* behaviors, some of which were demonstrated in Chapter 1.

Spectroscopic (Mizushima1954; Wilson 1959, 1962; Herschback 1963) and electron diffraction (Bonham and Bartell 1959; Kuchitsu 1959; Bartell and Kohl 1963) observations of small molecules with rotatable bonds has made it plain that rotation about sp^3-sp^3 C-C bonds is not uniform and without barriers. Rather, it has threefold symmetry with energy barriers separating energy minima for the staggered backbone conformations, as

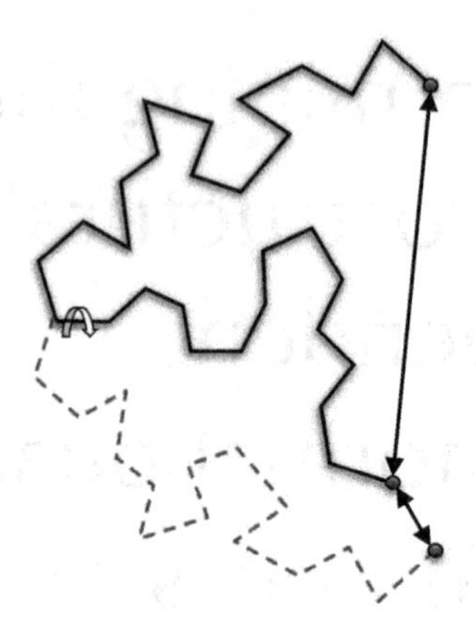

FIGURE 3.1 Illustration of the reduction in the end-to-end separation of polymer chain ends achieved by altering the conformation about a single backbone bond. (Reprinted with permission from Shen, J. and Tonelli, A. E., *J. Chem. Educ.*, 94, 1738–1745, 2017. Copyright 2017 American Chemical Society.)

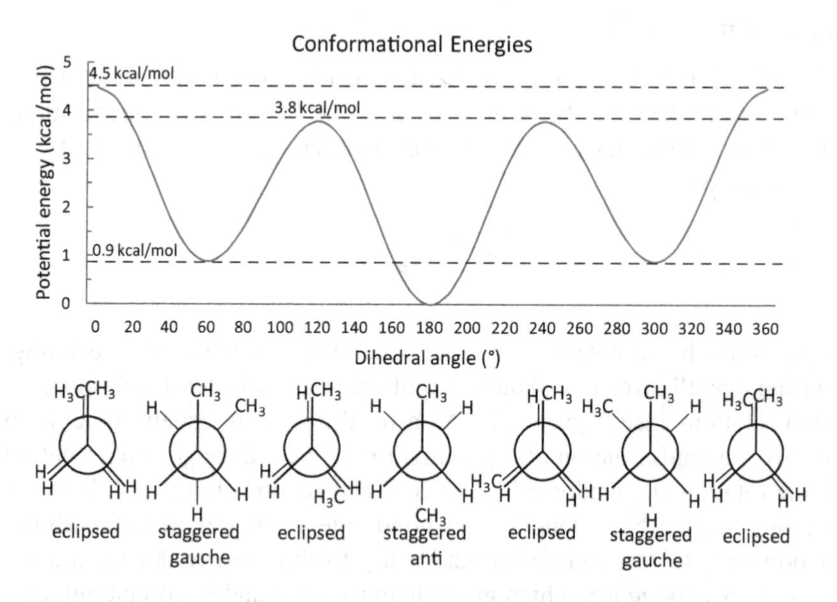

FIGURE 3.2 The conformations and energies for rotations about the central bond in butane. (Flory, P. J., *Statistical Mechanics of Chain Molecules*, Wiley-Interscience, New York, 1969.)

seen in Figure 3.2, which presents the rotational potential about the central bond in butane. The discrete nature of the rotational potentials about the backbone bonds in many polymers was recognized and the rotational isomeric states (RIS) approximation was adopted (Volkenstein 1963; Birshstein and Ptitsyn 1964). Each rotatable backbone bond was assumed to be able to adopt only a few low-energy rotational conformations or states usually selected to be coincident with rotational potential minima, as seen for butane in Figure 3.2.

Figure 3.3 presents drawings of some staggered pentane conformations and the conformational energy $E(\varphi_2,\varphi_3)$ contour map for pentane. Note that all low-energy regions correspond to both φ_2 and φ_3 assuming staggered conformations. However, when (φ_2,φ_3) are *gauche* ±, *gauche* ∓$(g\pm,g\mp=\pm120°,\mp120°)$, no energy minima appear, because of the close approach of methyl groups seen in drawing (c), which cause a repulsive interaction. This has been called a "pentane interference," which must also be present to some degree in all polymers, and demonstrates that polymer conformations are generally nearest neighbor dependent. Thus, the likelihood or probability of a bond rotation about a given bond depends on the rotational states of its nearest-neighbor backbone bonds. On the other hand, the locations of the low-energy RIS

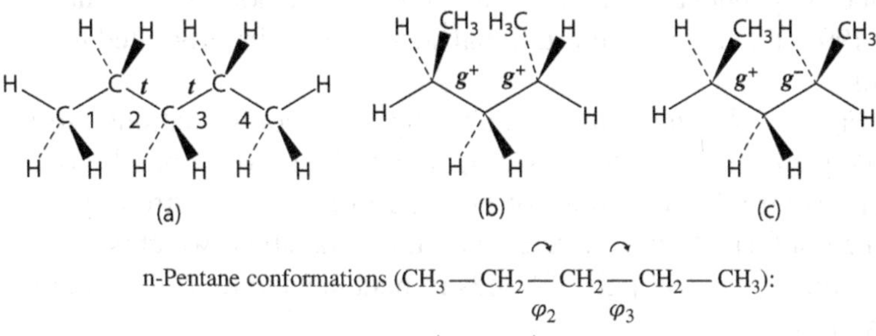

(a) *trans, trans* ($\varphi_2=\varphi_3=0°$); (b) *gauche*$^+$, *gauche*$^+$ ($\varphi_2=\varphi_3=120°$); and (c) *gauche*$^+$, *gauche*$^-$ ($\varphi_2=120°$, $\varphi_3=-120°$).

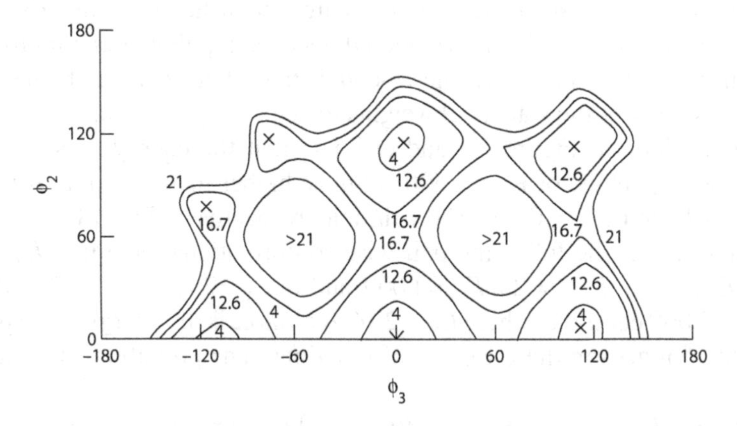

FIGURE 3.3 Pentane conformations (top) (a–c) and the pentane conformational energy map (bottom), showing only the top half. (Reprinted with permission from Abe, A. et al., *J. Am. Chem. Soc.*, 88, 631, 1966. Copyright 1966 American Chemical Society.)

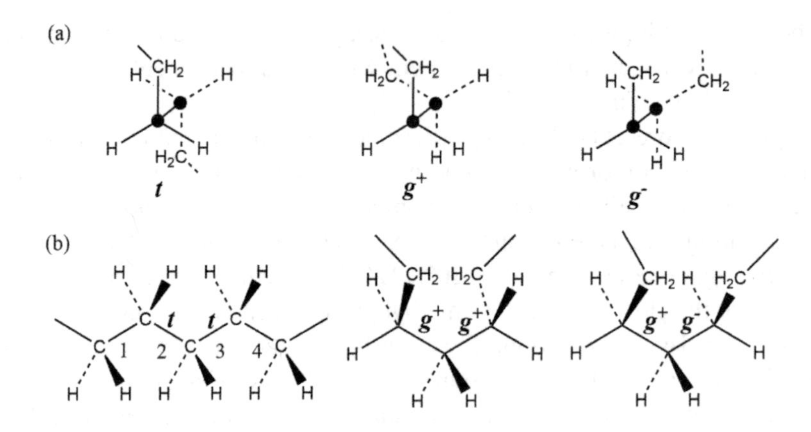

FIGURE 3.4 Single bond (a) and neighboring bond pair (b) PE conformations.

conformations do not generally depend on the conformations of the nearest neighbor bonds. For polymers whose RIS locations and energies are both neighbor dependent, it has been demonstrated (Flory 1969; Mattice and Suter 1994) that additional rotational states can be incorporated to account for this.

In Figure 3.4, single and neighboring bond conformations are shown for a Polyethylene (PE) chain. Comparison of the two-bond PE conformations in Figure 3.4 and those of pentane in Figure 3.3 strongly suggests that the conformations, energies, and their statistical weights are closely similar for the central pair of bonds in pentane and a neighboring pair of bonds in PE.

RIS models of polymer chains are most often established by using pair-wise statistical weights $e^{\frac{-E(\varphi_i,\varphi_{i+1})}{RT}}$ to represent nearest neighbor dependent conformations. Statistical weight values are established by comparison of observed and calculated conformational dependent polymer chain properties, such as the mean-square end-to-end distance between the chain ends, and adjustment of the statistical weights to achieve agreement.

We can look at Figure 3.2 and estimate that the energy of a *gauche* ($g^{\pm}$) bond in pentane or PE is ~500 cal/mole higher than that of a *trans* (t) bond, so E_t and $E_{g\pm} = 0$ and 500 cal/mole, respectively. From Figure 3.3, we can estimate the following pair-wise conformational energies: $E_{tt} = 0$, $E_{tg\pm} = E_{g\pm t} = 500$, $E_{g\pm g\pm} = 2 \times E_{g\pm} = 1{,}000$, and $E_{g\pm g\mp} = 2 \times E_{g\pm} + E_{\omega} = 2 \times 500 + 2{,}000 = 3{,}000$ cal/mole, where $E_{\omega} = 2{,}000$ cal/mole is the energy of the pentane interferences in the $g \pm g \mp$ conformations. The probabilities of these neighboring bond conformations, $P(\varphi_i, \varphi_{i+1})$ are simply $\frac{e^{\frac{-E(\varphi_i,\varphi_{i+1})}{RT}}}{\sum_{\varphi_i,\varphi_{i+1}} e^{\frac{-E(\varphi_i,\varphi_{i+1})}{RT}}}$, which leads to $P_{tt} = 0.322$, $P_{tg\pm} = P_{g\pm t} = 0.139$, $P_{g\pm g\pm} = 0.060$, and $E_{g\pm g\mp} = 0.001$, when T = 25°C.

The RIS conformational models of polymer chains are rigorous and have been successfully used to predict the conformational characteristics of individual polymer chains (Flory 1969; Tonelli 1986). Artificial models of polymer chains like freely jointed or freely or independently rotating chains and worm-like chains are often adopted because they have analytical expressions for their chain dimensions. This results in disregard of the distinguishing features of real polymer chains, i.e., their chain geometries, chemical structures, and resultant conformational preferences. These molecular features are critical to the behaviors and properties of polymer materials and are accounted for appropriately in the RIS conformational models of polymer chains. We now briefly describe how they may be established.

In the top part of Figure 3.5, a pentad stereosequence portion of a vinyl polymer chain is drawn with carbon atoms and bonds labeled. In the middle portion of the figure, Newman diagrams about bonds i and $i+1$ are drawn, with both in the *trans* conformation, and the bottom portion illustrates the pair-wise dependent φ_i, φ_{i+1} conformations. We now use these drawings to illustrate the development of an RIS model.

The substituted carbons are arbitrarily designated l or d if their attached side-chains R are, respectively, below or above the backbone plane. The 1st-order interactions dependent upon a single rotation angle φ_i or φ_{i+1} can be assigned from their Newman diagrams in Figure 3.5. For φ_i and $\varphi_{i+1} = t, g^+, g^-$ statistical weights $\eta, \tau, 1$ may be assigned. η represents a ratio of statistical weights for CH—R and CH—CH_2 interactions and τ accounts for the CH—R interaction. These are a result of CH—R, CH—R and CH—CH_2, and CH—CH_2 interactions occurring, respectively, when φ_i or $\varphi_{i+1} = t, g^+, g^-$.

In the bottom of Figure 3.5, the 2nd-order, pair-wise dependent, pentane-like interactions (ω_s) are illustrated for all nine $(\varphi_i, \varphi_{i+1})$ conformations. ω, ω', and ω'' correspond, respectively, to "pentane-like" interferences between CH_2—CH_2, R—CH_2, and R—R. Now, we collect these 1st- and 2nd-order statistical weights in a 3×3 matrix whose columns correspond to $\varphi_{i+1} = t, g^+, g^-$ and rows are designated by $\varphi_i = t, g^+, g^-$. It will soon be apparent why we do this. All elements in the 1st, 2nd, and 3rd columns, respectively, contain the 1st-order statistical weights η, τ, and 1 corresponding to $\varphi_{i+1} = t, g^+, g^-$. To these 1st-order statistical weights are appended the "pentane-like" interference, 2nd-order statistical weights illustrated and given in the bottom of Figure 3.5. Thus, we obtain:

$$
\begin{array}{cc}
 & \varphi_i/\varphi_{i+1} \quad\; t \quad\;\; g^+ \quad\; g^- \\
U_{ld} = & \begin{array}{c} t \\ g^+ \\ g^- \end{array} \begin{bmatrix} \eta & \tau\omega'' & \omega' \\ \eta\omega'' & \tau\omega'^2 & \omega \\ \eta\omega' & \tau\omega & 1 \end{bmatrix}.
\end{array}
$$

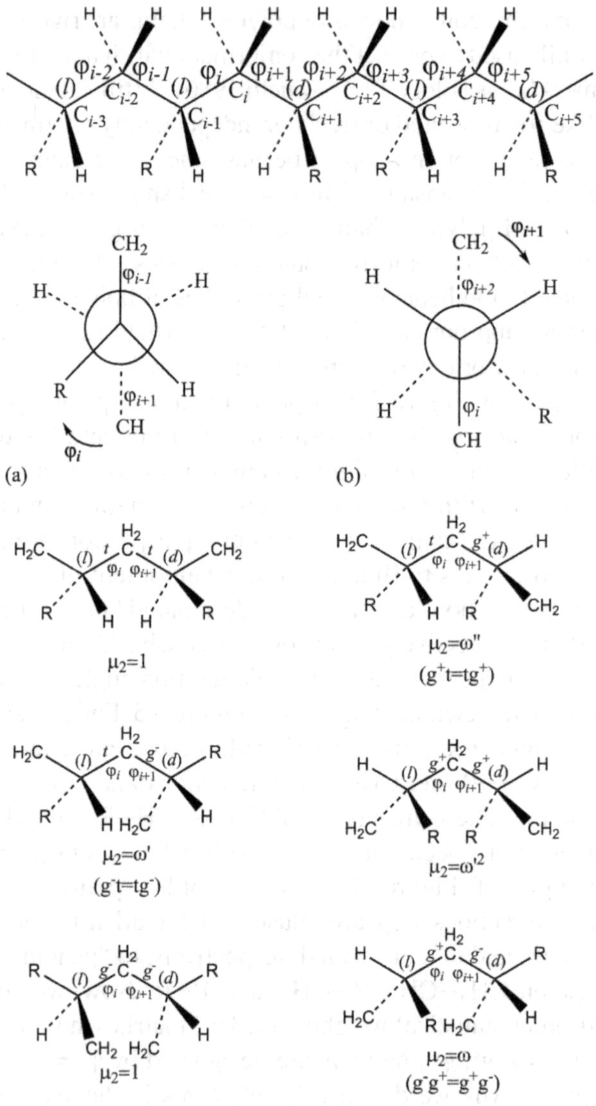

FIGURE 3.5 An mrrr vinyl polymer pentad stereosequence fragment (top), Newman diagrams along bonds i (a) and $i+1$ (b) (middle), and 2nd-order interactions dependent on φ_i and φ_{i+1} (bottom).

For the other racemic diad (dl) and the two meso diads (dd and ll), the following statistical weight matrices may be derived (Flory 1969; Bovey 1982):

$$U_{dl} = \begin{bmatrix} \eta & \omega' & \tau\omega'' \\ \eta\omega' & 1 & \tau\omega \\ \eta\omega'' & \omega & \tau\omega'^2 \end{bmatrix},$$

$$U_{ll} = \begin{bmatrix} \eta\omega'' & 1 & \tau\omega' \\ \eta\omega' & \omega' & \tau\omega\omega'' \\ \eta & \omega & \tau\omega' \end{bmatrix},$$

$$U_{dd} = \begin{bmatrix} \eta\omega'' & \tau\omega' & 1 \\ \eta & \tau\omega' & \omega \\ \eta\omega' & \tau\omega\omega'' & \omega' \end{bmatrix}.$$

For bond pairs flanking the asymmetric centers, two statistical weight matrices U_l and U_d can be similarly derived:

$$U_l = \begin{bmatrix} \eta & \tau & 1 \\ \eta & \tau & \omega \\ \eta & \tau\omega & 1 \end{bmatrix},$$

$$U_d = \begin{bmatrix} \eta & 1 & \tau \\ \eta & 1 & \tau\omega \\ \eta & \omega & \tau \end{bmatrix}.$$

The configurational partition function for that portion of an atactic vinyl polymer shown in Figure 3.5, which is a pentad stereosequence, becomes $Z = J^{*}U_lU_{ll}U_lU_{ld}U_dU_{dl}U_lU_{ld}U_dJ$. Below, we will discuss the partition function Z, which is a sum of the statistical weights of each polymer conformation or $SW_{conf} = \sum_{i=2}^{n-1} e^{\frac{-E(\varphi_i,\varphi_{i+1})}{RT}}$ and the row and column vectors J^* and J used in its evaluation.

The energy of any particular polymer chain conformation is given below, where $\{\varphi\}$ is the collection of consecutive backbone bond rotation angles $\varphi_2, \varphi_3, \varphi_4, \ldots, \varphi_{i-3}, \varphi_{i-2}, \varphi_{i-1}$ corresponding to that particular chain conformation.

$$E\{\varphi\} = \sum_{i=2}^{n-1} E_i\left(\varphi_{i-1}, \varphi_i\right) = \sum_{i=2}^{n-1} E_{\xi\eta;i},$$

where ξ and η denote the rotational states of bonds $i-1$ and i, respectively.

This means we can treat each distinct polymer chain conformation, whose energy and statistical weight or probability can be easily evaluated by summing its pair-wise dependent energies, $E(\varphi_{i-1}, \varphi_i)$, as a statistical mechanical system. The entire collection of chain conformations then makes up an easily treatable and rigorous statistical mechanical ensemble (Hill 1960).

Summing over all possible chain conformations leads formally to the conformational partition function Z or the sum of all SW_{conf}s, which is a measure of a polymers conformational flexibility, i.e., the larger the Z, the more conformations of comparable energy that are available to the polymer. Of course, the thermodynamic characteristics of polymer conformations can be evaluated from their partition functions (Hill 1960). We shall also soon see that we can utilize Z to obtain several conformationally averaged polymer chain properties, including chain dimensions and bond conformational populations. In many instances the INSIDE conformational characteristics of polymer chains, which are sensitively dependent on their microstructures, can be connected to and used to understand the OUTSIDE behaviors of their materials.

Statistical weights $\mu_{\xi\eta}$, or Boltzmann factors, corresponding to the energies $E_{\xi\eta}$ may be defined as:

$$\mu_{\xi\eta;i} = e^{\frac{-E_{\xi\eta;i}}{RT}},$$

and expressed in matrix form:

$$U_i = \left[\mu_{\xi\eta;i}\right] = \begin{bmatrix} \mu_{\alpha\alpha} & \mu_{\alpha\beta} & \cdots & \mu_{\alpha\nu} \\ \mu_{\beta\alpha} & \mu_{\beta\beta} & \cdots & \mu_{\beta\nu} \\ \vdots & \vdots & & \vdots \\ \mu_{\nu\alpha} & \mu_{\nu\beta} & \cdots & \mu_{\nu\nu} \end{bmatrix}.$$

The rows of the $\nu \times \nu$ matrix U_i are indexed with the states ξ of bond $i-1$ and columns with the states η of bond i.

The statistical weight of a particular chain conformation is then simply:

$$\Omega_{\{\varphi\}} = \prod_{i=2}^{n-1} \mu_{\xi\eta;i}$$

and the sum of the statistical weights of each conformation is the partition function Z:

$$Z = \sum_{\{\varphi\}} \Omega_{\{\varphi\}} = \sum_{\{\varphi\}} \prod_{i=2}^{n-1} \mu_{\xi\eta;i}.$$

Application (Flory 1969) of matrix methods (Kramers and Wannier 1941), which were previously used to treat the Ising ferromagnet (Ising 1925), leads to:

$$Z = J^* \left[\prod_{i=2}^{n-1} U_i \right] J,$$

where J^* and J are the $1 \times \nu$ and $\nu \times 1$ row and column vectors:

$$J^* = \begin{bmatrix} 1 & 0 & \cdots & 0 \end{bmatrix}, \quad J = \begin{bmatrix} 1 \\ 1 \\ \vdots \\ 1 \end{bmatrix}.$$

These mathematical matrix methods are applicable to linear polymers, because they are essentially 1-dimensional systems with overall conformational energies $E\{\varphi\}$ that are summations of their nearest neighbor dependent pair-wise interactions.

$$E\{\varphi\} = \sum_{i=2}^{n-1} E_i \left(\varphi_{i-1}, \varphi_i \right).$$

Using a very small example linear chain, n-hexane

$\left(CH_3 - CH_2 \overset{\curvearrowright}{\underset{\varphi_1}{-}} CH_2 \overset{\curvearrowright}{\underset{\varphi_2}{-}} CH_2 \overset{\curvearrowright}{\underset{\varphi_3}{-}} CH_2 - CH_3 \right)$, we now illustrate the means and the importance of determining the conformational partition function Z of a polymer. The conformations of n-hexane are 3 × 3 × 3 = 27 in number, and they along with their relative energies $E\left(\varphi_1, \varphi_2, \varphi_3\right)$ and statistical weights $\mu\left(\varphi_1, \varphi_2, \varphi_3\right)$ are presented in Table 3.1.

Boltzmann factors for the 1st- and 2nd-order interactions in n-alkanes and PE were evaluated at 25°C, i.e., $\sigma = e^{\frac{-500}{RT}} = 0.43$ and $\omega = e^{\frac{-2000}{RT}} = 0.034$. As an example, let's consider the $g^{\pm}g^{\pm}g^{\mp}$ n-hexane conformation containing three gauche bond conformations. The 1st-order *gauche* interactions contribute 3 × 500 = 1,500 cal to the total conformational energy, while the nearest neighbor pair $\varphi_2, \varphi_3 = g^{\pm}g^{\mp}$ produce a 2nd-order pentane interference and contribute an additional 2,000 cal. This results in $E\left(g^{\pm}g^{\pm}g^{\mp}\right) = 3500\,\text{cal}$, $\mu\left(g^{\pm}g^{\pm}g^{\mp}\right) = \sigma^3\omega = 0.003$, and a relative probability or population of 0.0005.

TABLE 3.1
Energies, Statistical Weights, and Probabilities of n-Hexane Conformations

Conformer			$E(\varphi_1,\varphi_2,\varphi_3)$		
φ_1	φ_2	φ_3	Kcal/mole	$\mu(\varphi_1,\varphi_2,\varphi_3)$	$P(\varphi_1,\varphi_2,\varphi_3)$
t	t	t	0.0	1.0	0.189
t	t	$g\pm$	0.5	$\sigma=0.43$	0.081
$g\pm$	t	t	0.5	$\sigma=0.43$	0.081
t	$g\pm$	t	0.5	$\sigma=0.43$	0.081
t	$g\pm$	$g\pm$	1.0	$\sigma^2=0.19$	0.036
$g\pm$	$g\pm$	t	1.0	$\sigma^2=0.19$	0.036
$g\pm$	t	$g\pm$	1.0	$\sigma^2=0.19$	0.036
$g\pm$	t	$g\mp$	1.0	$\sigma^2=0.19$	0.036
$g\pm$	$g\pm$	$g\pm$	1.5	$\sigma^3=0.08$	0.015
t	$g\pm$	$g\mp$	3.0	$\sigma^2\omega=0.006$	0.001
$g\pm$	$g\mp$	t	3.0	$\sigma^2\omega=0.006$	0.001
$g\pm$	$g\pm$	$g\mp$	3.5	$\sigma^3\omega=0.003$	0.0005
$g\mp$	$g\pm$	$g\pm$	3.5	$\sigma^3\omega=0.003$	0.0005
$g\pm$	$g\mp$	$g\pm$	5.5	$\sigma^3\omega^2=0.0$	0.0

$Z=\sum_{\varphi_1,\varphi_2,\varphi_3}\mu(\varphi_1,\varphi_2,\varphi_3)=1+6\sigma+8\sigma^2+4\sigma^2\omega+2\sigma^3+4\sigma^3\omega+2\sigma^3\omega^2=5.29$ and $P(\varphi_1,\varphi_2,\varphi_3)=\mu(\varphi_1,\varphi_2,\varphi_3)/Z$

The statistical weight matrices for n-hexane are, respectively,

$$U_1=\begin{bmatrix}1&\sigma&\sigma\\1&\sigma&\sigma\\1&\sigma&\sigma\end{bmatrix},\ U_2=U_3=\begin{bmatrix}1&\sigma&\sigma\\1&\sigma&\sigma\omega\\1&\sigma\omega&\sigma\end{bmatrix},$$

and its conformational partition function is given by:

$$Z=J^*\left[\prod_{i=2}^{n-1}U_i\right]J=J^*\left[U_1U_2U_3\right]J,$$

where J^* and J are the row and column vectors $\begin{bmatrix}1&0&0\end{bmatrix}$ and $\begin{bmatrix}1\\1\\1\end{bmatrix}$.

After multiplication of the statistical weight matrices, the partition function obtained is $Z=1+6\sigma+8\sigma^2+4\sigma^2\omega+2\sigma^3+4\sigma^3\omega+2\sigma^3\omega^2=5.29$, which not surprisingly is identical to the n-hexane partition function obtained "by hand" in Table 3.1. This exercise had the purpose of convincing you that multiplying the RIS statistical weight matrices does in fact give the correct conformational partition function.

We can also determine bond conformation populations using Z. For instance, if we sum up all the probabilities of conformations with $\varphi_2 = g^{\pm}$ from Table 3.1, we find a fractional population of 0.342. If one replaces the elements in the first (*t*) column of U_2 with 0_S to evaluate $Z(\varphi_2 = g^{\pm})$ and then divide it by the complete Z, i.e., $P(\varphi_2 = g^{\pm}) = Z(\varphi_2 = g^{\pm}) / Z$, it also yields $P(\varphi_2 = g^{\pm}) = 0.342$.

As an aside, it is ironic that many chemists consider polymers to be more complex and difficult to understand. However, and to the contrary, their conformational characteristics, which we are attempting to demonstrate in this book can be related to the behaviors of their materials, can be treated rigorously by statistical mechanics. On the other hand, the statistical mechanics of small molecule systems is beyond rigorous treatment, because estimating the energies of large collections of small molecules is an insoluble many-body problem.

The statistical weights of nearest neighbor pair-wise dependent polymer chain conformations depend on their microstructures. For example, the statistical weight matrices and conformational partition function for the example vinyl polymer in Figure 3.5 treated here depend on the chemical structure of its side-chains R, as well as their connectivity to the backbone or stereosequence. Before we can utilize the derived RIS statistical weight matrices, the pair-wise energies need to be determined so the statistical weight parameters (η and τ 1st-order and $\omega, \omega', \omega''$ 2nd-order) can be evaluated.

As an example of deriving the RIS conformational model for a polymer, we utilize the conformational analysis of poly(2-vinyl pyridine) (P2VP), which is shown below in Figure 3.6 (Tonelli 1985). To estimate the pairwise conformational energies $E(\varphi_1, \varphi_2)$, rotation about each backbone C—C bond was assigned a threefold intrinsic torsional potential with a barrier height of 2.8 kcal/mol. To evaluate non-bonded van der Waals and electrostatic interactions between the atoms in this P2VP fragment, a Lennard-Jones 6–12 potential function together with a Coulombic term were utilized (Yoon et al. 1975a; Tonelli 1982). Based on the dipole moment measured for pyridine in dilute solutions (2.25D, McClellan 1963), partial charges of −0.68 and +0.34 were assigned, respectively, to the N atom and the two C atoms bonded to it, and a dielectric constant of 3.5 was assumed to mediate these electrostatic interactions.

Rotation angles φ_1 and φ_2 were stepped in 10° increments, while the pyridine rings were limited to two orientations: $\chi = 0°$, where C_2—N and C_α—H_α are eclipsed and $\chi = 180°$, where C_2—C_3 and C_α—H_α are eclipsed (see Figure 3.6). For these two conformations, the pyridine ring plane bisects

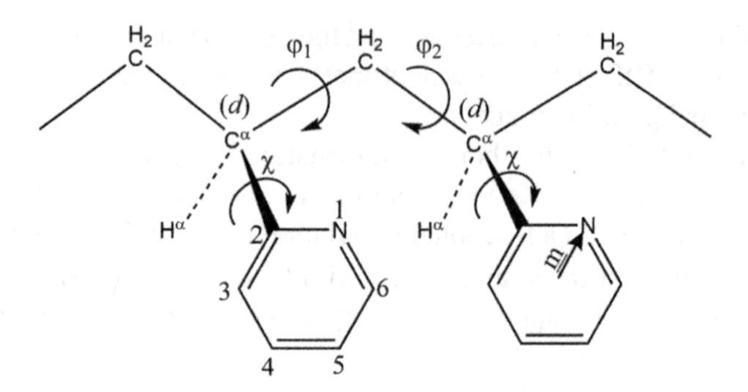

FIGURE 3.6 A five backbone carbon fragment of a meso (m) diad of P2VP.

the $\angle CH_2—C_\alpha H—CH_2$ valence angle, as suggested from previous studies indicating that only small deviations (10°–30°) from $\chi = 0°$ or 180°, are permitted by the resultant steric interactions (Abe et al. 1970; Tonelli 1973).

Vinyl polymers with large side-chain substituents in certain conformations can prevent solvent molecules from gaining access to the polymer backbone. However, when the *m* P2VP diad in Figure 3.6 adopts the tg^- conformation, the pyridine rings are sufficiently separated to permit access of the solvent. Yoon et al. (1975b) devised a procedure to account for the conformational dependence of solvent interactions. When the distance r between side-chains becomes sufficiently great ($r \geq \sigma$), they replaced side-chain-side-chain interactions with side-chain-solvent interactions. At the distance σ, the energy of interaction should level off and remain constant for all distances $r > \sigma$. Based on previous observations, σ = 4–5Å seems reasonable, so we used σ = 4.5Å and ∞ (no solvent-polymer interactions considered) in our calculations on P2VP (Sundararajan and Flory 1974; Sundararajan 1977a, 1977b, 1978, 1980; Yoon et al. 1977a, 1977b; Tonelli 1982).

The conformational energy map calculated as described for the *m* (*d*, *d*) diad of P2VP is presented in Figure 3.7. Similar conformational energy calculations were done for the *r* (*d*, *l*) P2VP diad. In both cases, the energy maps make apparent that backbone conformations that result in the simultaneous gauche arrangements of C_α, CH_2, and the pyridine side-chain (Pyr), as shown below in the upper part of Figure 3.7 for the *m* P2VP diad, are precluded. As a consequence, each backbone bond is essentially limited to two rotational states.

The statistical weight matrices U' and U'', for the bond pairs flanking and between substituted carbons can be written as,

$$U' = \begin{bmatrix} 1 & 1 \\ 1 & 0 \end{bmatrix} \quad U'' = \begin{bmatrix} Z_{tt} & Ztg^{\pm} \\ Zg^{\pm}t & Zg^{\pm}g^{\pm} \text{ or } Zg^{\pm}g^{\mp} \end{bmatrix},$$

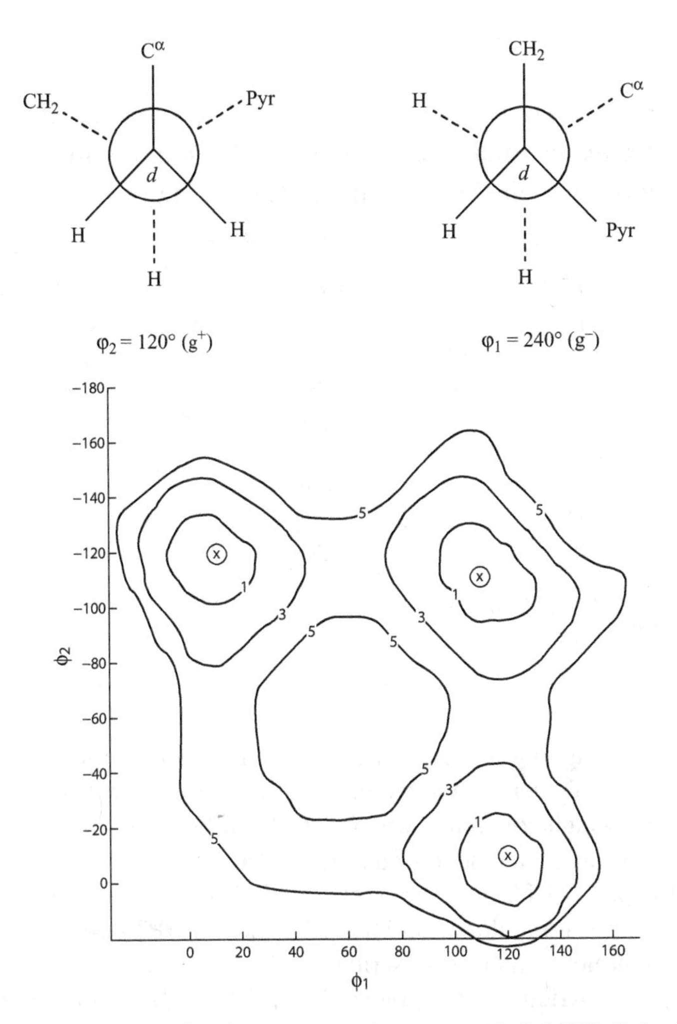

FIGURE 3.7 Conformational energy map for a meso (*d, d*) P2VP diad calculated with $\chi = 180°$ and $\sigma = 4.5$. X indicates the locations of the lowest energy conformations at $\varphi_1,\varphi_2 = 10°, -120°$; $120°, -10°$; and $110°, -110°$. The contours are drawn in units of kcal/mol relative to X. (Reprinted with permission from Tonelli, A. E., *Macromolecules*, 18, 2579, 1985. Copyright 1985 American Chemical Society.)

where Z_S are the partition functions of the low-energy conformational domains. Table 3.2 presents the Z_S of the low-energy conformational domains obtained from the conformational energy contour maps calculated for m(dd) and r(dl) P2VP diads.

We may also express the statistical weight matrices of P2VP in terms of the 1st- and 2nd-order interaction parameters dependent upon a single

TABLE 3.2
Partition Functions Z_s, Average Energies $<E_s>$, and Statistical Weight Pre-exponential Factors α_s for P2VP Diad Conformation Obtained from Their Conformational Energy Maps

	Z_s[a]		$<E_s>$, kcal/mol		a_s	
State(s)	$\sigma = 4.5$ Å	$\sigma = \infty$	$\sigma = 4.5$ Å	$\sigma = \infty$	$\sigma = 4.5$ Å	$\sigma = \infty$
d, d; tt	0.10	0.41	1.45	0.49	0.99	0.87
d, d; tg⁻	2.79	1.0	−0.73	0.03	0.90	0.96
d, d; g⁺t	2.79	1.0	−0.73	0.03	0.90	0.96
d,d; g⁺g⁻	3.77	0.58	−0.68	0.56	1.31	1.38
d, l; tt	1.0	1.0	0.0	0.0	1.0	1.0
d, l; tg⁺	0.44	0.11	0.76	1.78	1.44	1.74
d, l; g⁺t	0.44	0.11	0.76	1.78	1.44	1.74
d, l; g⁺g⁺	7.29	1.37	−0.98	0.09	1.59	1.58

Source: Tonelli, A. E., *Macromolecules*, 18, 2579, 1985.

[a] $Z_s = a_s e^{\frac{-<E_s>}{RT}}, T = 50°\text{C}$

or a pair of neighboring rotations, as was done above for our example generic vinyl polymer. Again, when the racemic (*dl* or *ld*) tt state is given a statistical weight of unity, we obtain the following solutions for the 1st- and 2nd-order interaction parameters from the results in Table 3.2: $\omega = 1.4\ \exp(-340/T)$, $\omega' = 1.6 \times \exp\text{-}(-820/T)$, $\omega'' = 0.9\ \exp(-E_{\omega''}/RT)$, where $E_{\omega''} = 0.5\text{–}1.5$ kcal/mol, and $\eta = 1.0 \times \exp(-E_\eta/RT)$, where $E_\eta = -0.1$ to $+0.8$ kcal/mol. η and ω'' describe the 1st- and 2nd-order interactions involving the pyridine side groups (Flory 1969). Both are sensitive to potential solvent interactions, leading to a range of possible values for E_η and $E_{\omega''}$. On the other hand, E_ω, and $E_{\omega'}$ are not sensitive to σ, because they characterize the 2nd-order CH_2—CH_2 and CH_2—Pyr interactions, which do not involve the close approach of two neighboring Pyr rings.

$$U'_d = \begin{array}{c} \\ t \\ g^- \end{array} \begin{array}{c} t \quad g^+ \\ \begin{bmatrix} 1 & 1 \\ 1 & 0 \end{bmatrix} \end{array} \quad U'_l = \begin{array}{c} \\ t \\ g^+ \end{array} \begin{array}{c} t \quad g^- \\ \begin{bmatrix} 1 & 1 \\ 1 & 0 \end{bmatrix} \end{array}$$

$$U''_{d,d} = \begin{array}{c} \\ t \\ g^+ \end{array} \begin{array}{c} t \qquad g^- \\ \begin{bmatrix} \omega'' & 1/\eta \\ 1/\eta & \omega/\eta^2 \end{bmatrix} \end{array} \quad U''_{l,l} = \begin{array}{c} \\ t \\ g^- \end{array} \begin{array}{c} t \qquad g^+ \\ \begin{bmatrix} \omega'' & 1/\eta \\ 1/\eta & \omega/\eta^2 \end{bmatrix} \end{array}$$

$$U''_{d,l} = \begin{matrix} & t & g^+ \\ t & 1 & \omega'/\eta \\ g^+ & \omega'/\eta & 1/\eta^2 \end{matrix} \quad U''_{l,d} = \begin{matrix} & t & g^- \\ t & 1 & \omega'/\eta \\ g^- & \omega'/\eta & 1/\eta^2 \end{matrix}.$$

Pair-wise conformational populations for P2VP *m* (*r*) diads obtained from its RIS model as embodied in the above statistical weight matrices are P_{tt} = 0.074 (0.248), $P_{tg^+} = P_{g^+t}$ = 0.325 (0.045), $P_{g^+g^-}$ = 0.726 (*m*), and $P_{g^+g^+}$ = 0.662 (*r*). Notice how different they are from the pair-wise conformer populations of PE and how sensitive they are to the stereosequences of P2VP.

In addition to polymer bond conformational populations, RIS conformational models (*U*s) and matrix multiplication can also be utilized to calculate their dimensions and other conformationally sensitive properties. Below we show, schematically, an all carbon backbone polymer with n bonds of length *l*, and valence angles $\pi - \theta$. We wish to calculate its mean-square end-to-end C_o to C_n length ($<r^2>$) averaged over all of its conformations.

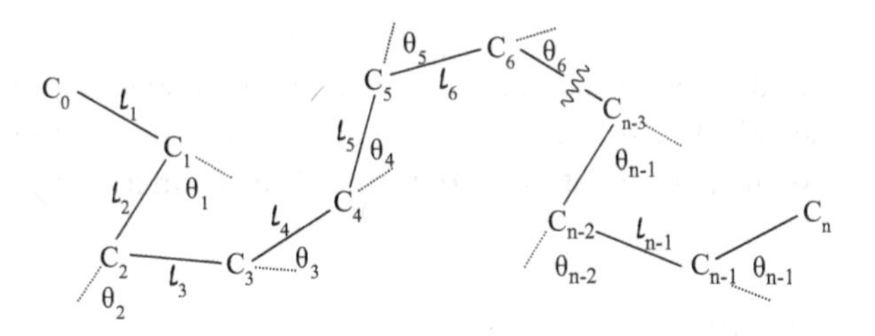

In theory, this can be achieved by using the Euler transformation matrix which can transforms a vector from a bond l_{i+1} to the previous bond l_i (Eyring 1932; Oka 1942; Flory 1969). If all n – 1 backbone bonds are transformed to the coordinate system along l_1 they may be added, and the squared sum of all bond x,y,z components would yield r^2.

$$T_i = \begin{bmatrix} \cos\theta_i & \sin\theta_i & 0 \\ \sin\theta_i \cos\varphi_i & -\cos\theta_i \cos\varphi_i & \sin\varphi_i \\ \sin\theta_i \sin\varphi_i & -\cos\theta_i \sin\varphi_i & -\cos\varphi_i \end{bmatrix}$$

If this procedure were repeated for each conformation and each resultant r^2 was multiplied by its probability, their sum would yield the conformationally averaged dimensions ($<r^2>$). However, there are x^n total conformations for a polymer chain of n bonds, each capable of assuming x distinct conformations. For a 10,000 bond [molecular weight (MW)] = 140,000 Polyethylene (PE) chain, this would be $3^{10,000} = 10^{4,800}$ conformations, and so this "brute force" method is not practical for obtaining average polymer chain dimensions.

Flory and Jernigan have shown that the average dimensions ($<r^2>_o$) can be readily obtained using polymer RIS statistical weight matrices U, bond lengths and valence angles (l and $\pi - \theta$), and the Euler matrix $T_i(\pi - \theta_i, \varphi_i)$. Even the dimensions of the polymer chains containing many thousands of backbone bonds may be quickly obtained by their computationally efficient matrix multiplication method.

$$\langle r^2 \rangle_0 = 2Z^{-1}K^* \left(\prod_{i-1}^{n} G_i \right) K,$$

where

$$G_i = \begin{bmatrix} U_i & \left(U_i \otimes l_i^T\right)\|T\| & \left(l_i^2/2\right)U_i \\ 0 & \left(U_i \otimes E\right)\|T\| & U_i \otimes l_i \\ 0 & 0 & U_i \end{bmatrix}$$

and l_i and l_i^T are the bond vectors in column and row representation, $l_i^2 = |l_i|^2$, E is the v order identity matrix, v is the number of bond rotational states, $\|T\|$ is the diagonal representation of the Euler transformation matrix T_i,

$$\|T_i\| = \begin{bmatrix} T(\varphi_i = \alpha) & & & \\ & T(\varphi_i = \beta) & & \\ & & \ddots & \\ & & & T(\varphi_i = v) \end{bmatrix},$$

and $\otimes$ denotes the direct product of two matrices (Flory 1969). $K^* = \left[J^* 00\ldots0\right](1 \times 5v)$,

$$K = \begin{bmatrix} 0 \\ 0 \\ \vdots \\ \vdots \\ 0 \\ J \end{bmatrix} (5 \times 1),$$

and the $(1 \times \nu)$ row and $(\nu \times 1)$ column vectors J^* and J have been described above.

In the Appendix at the end of this chapter, Fortran programs for calculating polymer chain dimensions are presented and explained.

RIS conformational models in combination with the matrix multiplication techniques just described can be used to estimate virtually any conformationally sensitive polymer chain property if, in addition to the polymer's RIS model (*U*s) and geometry (*l*s and valence angles $= \pi - \theta$), the contribution made by each bond or repeat unit of the polymer to the particular property is known. These properties include dipole moments, optical anisotropy (depolarization of scattered light), electrical-birefringence (Kerr effect), optical rotatory power, equilibrium populations of cyclic and linear chains and polymer stereosequences, as well as dimensions, and conformational energies, entropies, and populations (Flory 1969). We will be presenting examples of how these estimated conformation-dependent properties have been successfully correlated with their experimental values. It should be mentioned that their temperature dependencies may also be estimated and have been used to help validate the RIS models of several polymers (Flory 1969; Mattice and Suter 1994).

We now revisit P2VP and use these matrix methods to calculate its expected dimensions, dipole moments, and equilibrium stereosequences for comparison to their observed values in order to assess the validity of the RIS model developed above for this vinyl polymer (Tonelli 1985). Using the statistical weight matrices ($U'_{d,l}$ and $U''_{dd,\ ll,\ dl,\ ld}$) derived from the conformational energy diagrams for P2VP (see Figure 3.7 and Table 3.2) and the aforementioned matrix multiplication methods, we calculated the mean-square end-to-end distances and dipole moments, $\langle r^2 \rangle_o$ and $\langle \mu^2 \rangle_o$, of P2VP chains of different tacticities, assuming a random or Bernoullian distribution of *meso* and *racemic* diads. 1st- and 2nd-order interaction energies $E_\eta = -0.1$ kcal/mol and $E_{\omega''} = 0.9$ kcal/mol were used. Instead of using the bond vectors *l*, the dipole moment vectors *m* of all P2VP repeat units were summed and averaged over all conformations to obtain $\langle \mu^2 \rangle_o$. The dipole moment (m = 2.25 D) measured for pyridine (McClellan 1963), as located in Figure 3.6, was transformed from the pyridine ring to the usual coordinate system along the C_α—CH_2 backbone bond to the right of the C_α carbon to which the pyridine ring is attached resulting in:

$$m = \begin{bmatrix} 0.794 \\ -1.179 \\ \pm 0.374\ d,l \end{bmatrix} D(50°C)$$

TABLE 3.3
$C_m = \langle\mu^2\rangle_o/x$ and $C_r = \langle r^2\rangle_o/nl^2$ Calculated for 400 Bond P2VP Chains with Bernoullian Probabilities P_m for Meso (*m*) Diads[a]

	C_m		C_r	
P_m	0°C	50°C	0°C	50°C
0.0	3.06	2.93	11.2	11.1
0.1	3.10	2.98	9.74	9.66
0.2	3.07	2.96	8.50	8.43
0.3	2.99	2.97	7.82	7.72
0.4	2.93	2.85	7.39	7.27
0.5	2.89	2.77	7.18	7.01
0.6	2.77	2.72	7.18	6.95
0.7	2.74	2.70	7.39	7.07
0.8	2.74	2.68	7.65	7.23
0.9	2.82	2.73	8.29	7.68
1.0	2.97	2.83	9.08	8.23

[a] $E_\eta = -0.1$ kcal/mol and $E_{\omega''} = 0.9$ kcal/mol. For $P_m = 0.1 - 0.9$ ten Monte Carlo chains were generated and C_m and C_r were averaged over this ensemble.

and which replaces the bond vector l in their matrix equation for $\langle r^2\rangle_o$. Table 3.3 presents the dimensions ($C_r = \langle r^2\rangle_o/nl^2$) and dipole moments per repeat unit x ($C_m = \langle\mu^2\rangle_o/x$) that were calculated at 0 and 50°C for P2VP chains of various tacticities (Tonelli 1985).

It is clear that the calculated dipole moments (C_ms) are virtually independent of P2VP stereosequences. To confirm this unexpected result, we measured the dipole moments of three P2VP samples with different stereosequences, and the results are seen in Table 3.4 (Tonelli 1985). Within experimental error, the observed squared dipole moments per repeat unit are ~3 D^2, regardless of stereosequence, and are in close agreement with the calculated values. Unlike P2VP, two other vinyl polymers poly(*p*-chlorostyrene) (Saiz et al. 1977) and poly(*N*-vinyl pyrrolidone) (Tonelli 1982) evidenced dipole moments that were sensitive to their stereosequences. While for P2VP, $C_m = 2.9 \pm 0.2$ D^2 regardless of tacticity, isotactic (all m diads or P_m ~1) poly(*p*-chlorostyrene) and poly(*N*-vinyl pyrrolidone) chains have predicted C_ms = 1.5, 3–5 times larger than those calculated for their syndiotacic and atactic chains.

Using the RIS model developed for P2VP chains with atactic stereosequences yields $C_r = \langle r^2\rangle_o/nl^2 = 7$–8 irrespective of the Bernoullian or

TABLE 3.4
Measured Dipole Moments for P2VPs with Different Stereosequences

Sample	Stereoregularity	$<\mu^2>/x$, D^2
A-1	Atactic, Bernoullian P2VP with $P_m = 0.5$	3.1[a]
A-2	Atactic, non-Bernoullian P2VP with $P_{mm} = 0.44$, $P_{mr} = P_{rm} = 0.18$, and $P_{rr} = 0.20$	2.8[a]
I	Isotactic, with P_{mm}~0.9	3.3[b]

[a] Measured in benzene at 25°C.

[b] Estimated by comparing the Dipolmeter readings obtained for 1.7 wt% solutions of samples I and A-2 in 1,4-dichlorobenzene at 80°C.

non-Bernoullian natures of their stereosequences and fall in the experimentally observed range of C_r = 7–10 (Arichi 1966; Dondos 1970). As can be seen from Table 3.3, in the range of stereosequences characterized by P_m = 0.2–0.8, the calculated dimensions, like the dipole moments, are insensitive to P2VP stereosequence. Only the dimensions of the highly syndiotactic (P_m = 0.0–0.1) and highly isotactic (P_m = 0.9–1.0) P2VP chains show modest predicted increases over those expected and observed for atactic P2VPs.

An additional conformation-dependent chain property that has been observed for P2VP oligomers is their equilibrium populations of stereosequences, as obtained from epimerization experiments. When the α-H is removed reversibly from the substituted $-CH_\alpha R-$ carbon in a vinyl polymer, generally requiring a catalyst, its stereochemical configuration may be altered or epimerized ($d \leftrightarrow l$). As time proceeds, eventually an equilibrium stereosequence will be produced, which is independent of the vinyl polymer's initial stereosequence.

Hogen-Esch and Tien and Hwang et al. have carried out such epimerizations on the oligomeric dimers, trimers, and tetramers of P2VP, with results presented in Table 3.5 (Hogen-Esch and Tien 1980; Hwang et al. 1981). Also presented there are the expected equilibrium stereosequence populations calculated for each of the P2VP oligomers. These were obtained by calculating the conformational partition functions (*Z*) of each stereosequence and dividing each by the sum of *Z*s for all stereosequences (Flory 1967; Suter 1981). Excellent agreement was obtained between the calculated and observed equilibrium stereoisomer fractions.

TABLE 3.5
Isomer Fractions *f* for P2VP Dimer, Trimer, and Tetramer Oligomers

	f	
Oligomer (isomer)	**Calculated**[a]	**Observed**[b]
dimer (m)	0.52	0.49
dimer (r)	0.48	0.51
trimer (mm)	0.24	0.23
trimer (mr + rm)	0.51	0.50
trimer (rr)	0.25	0.27
tetramer (mmm)	0.10	0.09
tetramer (mmr + rmm)	0.25	0.25
tetramer (mrm)	0.14	0.13
tetramer (rrm + mrr)	0.26	0.25
tetramer (rmr)	0.13	0.14
tetramer (rrr)	0.12	0.14

[a] T = 25°C.
[b] Measured by gas chromatography after epimerization at 25°C in t-BuOK/Me_2S0 for 150–300 h.

This agreement is particularly relevant as a test of the P2VP RIS conformational model we developed, because the calculated stereoisomer populations depend only on the statistical weight matrices. Unlike dimensions and dipole moments, they do not depend on the chain geometry (bond lengths and valence and/or the angles of the rotational states).

Though the agreement achieved between the predicted and observed conformation-dependent properties of P2VP is excellent and was achieved without modification of the conformational energies crudely estimated using semi-empirical potential functions, this is very seldom the case for polymer RIS conformational models. These estimated conformational energies are generally useful for defining the locations of the backbone rotational states, but more often than not, the energy estimates are just that, estimates, and their values and associated statistical weights must be treated as adjustable parameters. When several observed conformation-dependent properties are successfully reproduced by the same set of conformational statistical weights, then we can become confident of the RIS model used to obtain them.

The aliphatic polyester poly(D-α-hydroxybutyrate) (PHB) shown in Figure 3.8 was first isolated from bacteria (Lemoigne 1927), where it functions as a stored energy reserve (Doudoroff and Stanier 1959; Wilkinson 1963; Macrae and Wilkinson 1958; Williamson and Wilkinson 1958). RIS

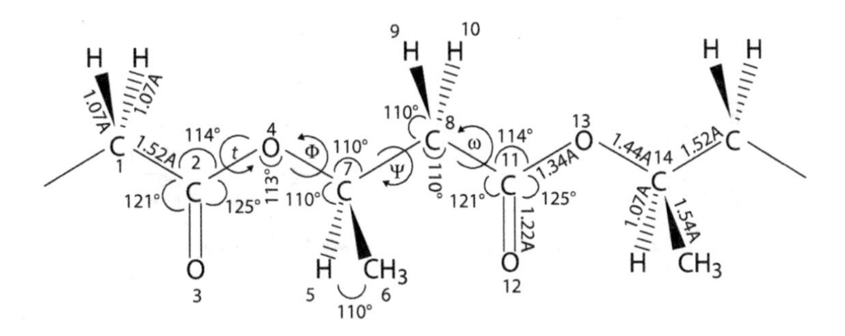

FIGURE 3.8 Schematic drawing of PHB showing structural parameters adopted for the conformational energy calculations used to develop the RIS model. (Reprinted with permission from Kyles, R.E. and Tonelli, A.E., *Macromolecules*, 36, 1125, 2003. Copyright 2003 American Chemical Society.)

conformational models were developed for PHB (Kyles and Tonelli 2003), as described earlier for P2VP, and from molecular dynamics simulation of isolated PHB chains (Hedenqvist et al. 1998).

The molecular dynamics simulation of isolated PHB chains is governed exclusively by intramolecular energetic interactions like the usual RIS treatment, but the interactions may be longer in range, i.e., between atoms further removed along the chain contour. Regardless of their separation along the chain contour, when a polymer chain conformation places segments within a specified through-space cutoff distance, their interaction is included in the energetic calculation.

The *Discover_3* module of the Biosym/MSI (San Diego, CA) software package was employed, and as in the traditional RIS energy calculations, $\varepsilon = 3$ was selected, and electrostatic and Van der Waals potentials were set to cut off at a distance >5.0 Å.

The bond conformation populations obtained for single isolated 10 and 60 repeat units PHB chain *via* molecular dynamics simulations compared to those obtained from the traditional RIS model are shown in Figure 3.9. Characteristic ratios of PHB dimensions obtained from the traditional RIS-derived model are C_∞ ~4 and roughly half of those $C_\infty = 8$ obtained from the RIS model which is obtained from the molecular dynamics simulation of isolated single PHB chains. The dimensions obtained from the molecular dynamics simulation of isolated single PHB chains are in agreement with the measured value (Aikita 1976; Miyaki et al. 1977; Kurata and Tsunashima 1989; Huglin and Radwan 1991).

Apparently intramolecular interactions longer than those dependent only on neighboring pairs of backbone rotations accounted for in the

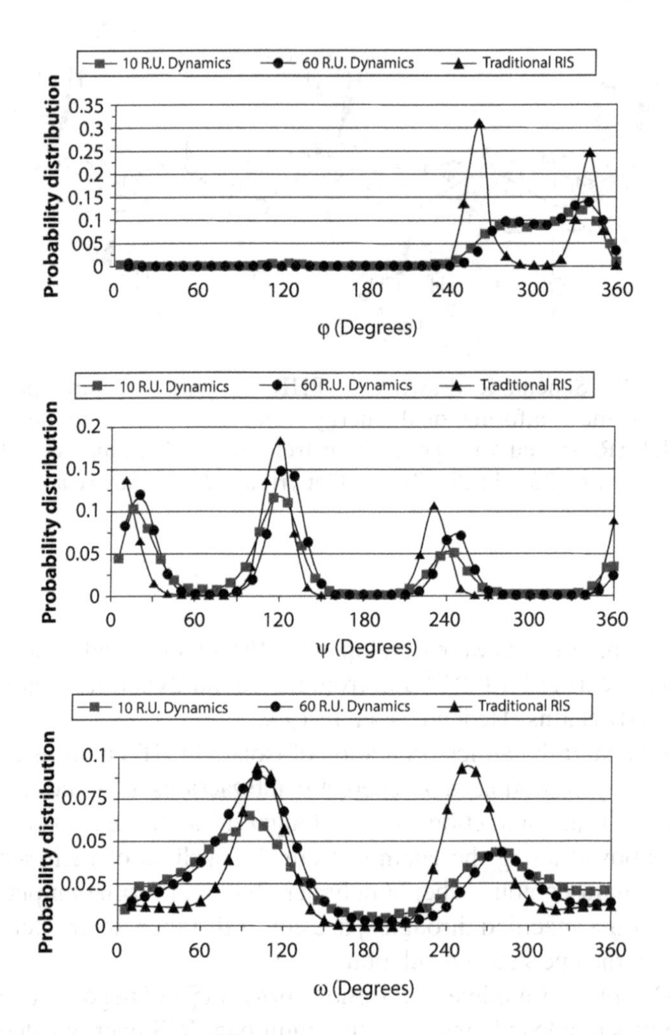

FIGURE 3.9 Comparison of the probability distributions for PHB backbone rotations obtained from the traditional RIS model and the single chain dynamics simulations of the PHB 10- and 60-mers. (Reprinted with permission from Kyles, R.E. and Tonelli, A.E., *Macromolecules* 36, 1125, 2003. Copyright 2003 American Chemical Society.)

traditional RIS model occur in PHB. More extended conformations are favored when single isolated PHB chain conformations are modeled by molecular dynamics simulation.

Finally, high-level quantum mechanical energy calculations have begun to be used to develop polymer RIS models. For example, conformational energies were calculated at the Second Order Møller–Plesset Perturbation Theory (MP2) level with a D95+(2df, p) basis set for 1,3-dimethoxy-propane as a model compound for poly-(oxytrimethylene) (Smith et al. 1996).

The 1st-order interaction energy $E(\varphi)$ for $\mathrm{O{-}C\overset{\curvearrowright}{\underset{\varphi}{-}}C{-}C}$ and 2nd-order interaction energy $E(\varphi_1,\varphi_2)$ for $\mathrm{O{-}C\overset{\curvearrowright}{\underset{\varphi_1}{-}}C\overset{\curvearrowright}{\underset{\varphi_2}{-}}C{-}O}$ were calculated to be –1.0 and 4.0 kcal/mol, respectively. However, these conformational energies did not produce statistical weights and U matrices that led to successful prediction of conformationally averaged and observed poly-(oxytrimethylene) dimensions and dipole moments. Carbone et al. employed *ab initio* Density Functional Theory (DFT) calculations to derive an RIS model for propene-norbornene copolymers, which was used with some success to conformationally average ^{13}C-Nuclear Magnetic Resonance (^{13}C-NMR) chemical shifts calculated using gauge including atomic orbitals with a large basis set (Carbone et al. 2003). At present, however, it appears that, like semi-empirically derived RIS models, RIS conformational models derived by quantum mechanical methods for polymers still need to be adjusted, calibrated, and validated by comparing conformationally averaged properties obtained with them to those that are observed experimentally.

REFERENCES

Abe, A., Jernigan, R. L., Flory, P. J. (1966), *J. Am. Chem. Soc.*, 88, 631.

Abe, Y., Tonelli, A. E., Flory, P. J. (1970), *Macromolecules*, 3, 294.

Aikita, S., Einaga, Y., Miyaki, Y., Fujita, H. (1976), *Macromolecules*, 9, 774.

Arichi, S. (1966), *Bull. Chem. Soc. Jpn.*, 39, 439.

Bartell, L. S., Kohl, D. A. (1963), *J. Chem. Phys.*, 39, 3097.

Birshstein, T. M., Ptitsyn, O. B. (1964), *Conformations of Macromolecules*, translated by Timasheff, S. N. and Timasheff, N. J. from the 1964 Russian Ed., Wiley-Interscience, New York.

Bonham, R. A., Bartell, L. S. (1959), *J. Am. Chem. Soc.*, 81, 3491.

Bovey, F. A. (1982), *Chain Structure and Conformations of Macromolecules*, Academic Press, New York, Chap. 7.

Carbone, P., Ragazzi, M., Tritto, I., Boggioni, L., Ferro, D. R. (2003), *Macromolecules*, 36, 891.

Dondos, A. (1970), *Makromol. Chem.*, 135, 181.

Doudoroff, M., Stanier, R. Y. (1959), *Nature*, 183, 1440.

Eyring, H. (1932), *Phys. Rev.*, 39, 746.

Flory, P. J. (1967), *J. Am. Chem. Soc.*, 89, 1798.

Flory, P. J. (1969), *Statistical Mechanics of Chain Molecules*, Wiley-Interscience, New York.

Hedenqvist, M. S., Bharadwaj, R., Boyd, R. H. (1998), *Macromolecules*, 31, 1556.

Herschback, D. R. (1963), *International Symposium on Molecular Structure and Spectroscopy*, Tokyo, 1962, Butterworths, London, UK.

Hill, T. L. (1960), *An Introduction to Statistical Mechanics*, Addison-Wesley, Reading, MA.

Hogen-Esch, T. E., Tien, C. F. (1980), *Macromolecules*, 13, 207.

Huglin, M. B., Radwan, M. A. (1991), *Polymer*, 32, 1293.

Hwang, S. S., Mathis, C., Hogen-Esch, T. E. (1981), *Macromolecules*, 14, 1802.
Ising, E. (1925), *Z. Phys.*, 31, 253.
Kramers, H. A., Wannier, G. H. (1941), *Phys. Rev.*, 60, 252.
Kuchitsu, K. (1959), *J. Chem. Soc. Jpn.*, 32, 748.
Kurata, M., Tsunashima, Y. In *Polymer Handbook*, 3rd ed.; Brandup, J., Immergut, E. H., Eds.; Wiley: New York, 1989; Chapter VII.
Kyles, R. E., Tonelli, A. E. (2003), *Macromolecules*, 36, 1125.
Lemoigne, M. (1927), *Ann. Inst. Pasteur.*, 41, 146.
Macrae, R. M., Wilkinson, J. F. (1958), *J. Gen. Microbiol.*, 19, 210.
Mattice, W. L., Suter, U. W. (1994), *Conformational Theory of Large Molecules*, Wiley-Interscience, New York.
McClellan, A. L. (1963), *Tables of Experimental Dipole Moments*, W. H. Freeman, San Francisco, CA.
Miyaki, Y., Einaga, Y., Hirosye, T., Fujita, H. (1977), *Macromolecules*, 10, 1356.
Mizushima, S. (1954), *Structure of Molecules and Internal Rotation*, Academic Press, New York.
Oka, S. (1942), *Proc. Phys. Math. Soc. Jpn.*, 24, 657.
Saiz, E., Mark, J. E., Flory, P. J. (1977), *Macromolecules*, 10, 967.
Smith, G. D., Jaffe, R. L., Yoon, D. Y. (1996), *J. Phys. Chem.*, 100, 13439.
Sundararajan, P. R. (1977a), *J. Polym. Sci., Polym. Lett. Ed.*, 15, 699.
Sundararajan, P. R. (1977b), *Macromolecules*, 10, 623.
Sundararajan, P. R. (1978), *Macromolecules*, 11, 256.
Sundararajan, P. R. (1980), *Macromolecules*, 13, 512.
Sundararajan, P. R., Flory, P. J. (1974), *J. Am. Chem. Soc.*, 96, 5025.
Suter, U. W. (1981), *Macromolecules*, 14, 523.
Tonelli, A. E. (1973), *Macromolecules*, 6, 682.
Tonelli, A. E. (1982), *Polymer*, 23, 676.
Tonelli, A. E. (1985), *Macromolecules*, 18, 2579.
Tonelli, A. E. (1986), *Encyclopedia of Polymer Science and Engineering*, 2nd ed., Wiley, New York, Vol. 4, p. 120.
Volkenstein, M. V. (1963), *Configurational Statistics of Polymer Chains*, translated by Timasheff, S. N. and Timasheff, N. J. from the 1964 Russian Ed., Wiley-Interscience, New York.
Wilkinson, J. F. (1963), J. Gen. *Microbiol.*, 32, 171.
Williamson, D. H., Wilkinson, J. F. (1958), *J. Gen. Micrbiol.*, 19, 198.
Wilson, E. B., Jr. (1959), *Ad. Chem. Phys.*, 2, 637.
Wilson, E. B., Jr. (1962), *Pure Appl. Chem.*, 4, 1.
Yoon, D. Y., Sundararajan, P. R., Flory, P. J. (1975a), *Macromolecules*, 8, 776.
Yoon, D. Y., Suter, U. W., Sundararajan, P. R., Flory, P. J. (1975b), *Macromolecules*, 8, 284.

DISCUSSION QUESTIONS

1. What characteristic attributes of polymer chains make it possible to rigorously treat their conformational characteristics by development of an RIS model and employing matrix multiplication techniques?
2. How can RIS conformational models be experimentally verified?

3. Can present state-of-the-art quantum mechanical methods be reliably used to derive RIS conformational models for polymers? Explain your answer.
4. Are the conformational energies and their statistical weights associated with RIS conformational models more independently verified by comparison between the observed populations of epimerized stereosequences and those predicted, or by calculated and observed chain dimensions $<r_2>_o$?

APPENDIX 3.1: FORTRAN PROGRAM FOR HEXANE "BY-HAND" CONFORMATIONAL POPULATIONS AND DISTANCES

```
     → Declaration of variables
real ph2,ph3,sph2,cph2,sph3,cph3,tt(3,3),t2(3,3),t3(3,3),
     7lcc(3,1),h1(3,1),tl(3,1),ttl(3,1),tttl(3,1),
     h2(3,1), h3(3,1),
     7u1(3,3),rt,z,udsum,dc16,prob,sigma,up(3,3),
     u2(3,3), v(3,3),
     7omega, tt2(3,3),tt23(3,3),u3(3,3),ph4,h4(3,1),
     sph4,cph4,
     7tt234(3,3),ttttl(3,1),t4(3,3),ud2sum, u4(3,3),u5(3,3)
     integer i,j,k
     → Output formats
  100 format(2f10.6)
  101 format(3f10.6)
     → Elements of transformation matrices
     tt(1,1)=0.37461
     tt(1,2)=0.92718
     tt(1,3)=0.0
     tt(2,1)=0.92718
     tt(2,2)=-0.37461
     tt(2,3)=0.0
     tt(3,1)=0.0
     tt(3,2)=0.0
     tt(3,3)=-1.0
     t2(1,1)=0.37461
     t2(1,2)=0.92718
     t2(1,3)=0.0
     t3(1,1)=0.37461
     t3(1,2)=0.92718
     t3(1,3)=0.0
     t4(1,1)=0.37461
     t4(1,2)=0.92718
     t4(1,3)=0.0
     → RT
```

```
rt=1.9872*298.16
→ Stat. weights, and U matrices
sigma=1.0/(2.71828**(500.0/rt))
omega=1.0/(2.71828**(2000.0/rt))
u2(1,1)=omega
u2(1,2)=sigma*omega
u2(1,3)=1.0
u2(2,1)=1.0
u2(2,2)=sigma*omega
u2(2,3)=omega
u2(3,1)=omega
u2(3,2)=sigma*omega*omega
u2(3,3)=omega
u3(1,1)=1.0
u3(1,2)=1.0
u3(1,3)=sigma
u3(2,1)=1.0
u3(2,2)=1.0
u3(2,3)=sigma*omega
u3(3,1)=1.0
u3(3,2)=omega
u3(3,3)=sigma
u1(1,1)=1.0
u1(1,2)=1.0
u1(1,3)=sigma
u1(2,1)=1.0
u1(2,2)=1.0
u1(2,3)=sigma
u1(3,1)=1.0
u1(3,2)=1.0
u1(3,3)=sigma
v(1,1)=0.0
v(1,2)=0.0
v(1,3)=0.0
v(2,1)=0.0
v(2,2)=0.0
v(2,3)=0.0
v(3,1)=0.0
v(3,2)=0.0
v(3,3)=0.0
→ Matrix multiplications
call mp(3,3,3,u1,u2,u4)
call mp(3,3,3,u4,u3,u5)
→ Partition function
z=u5(1,1)+u5(1,2)+u5(1,3)
→ C-C bond vector
lcc(1,1)=1.53
```

```
lcc(2,1)=0.0
lcc(3,1)=0.0
→ Transformation of backbone bonds to ref. frame
  of 1st bond for all conformations
call mp(3,3,1,tt,lcc,tl)
call add(3,lcc,tl,h1)
→ Stepping each backbone rotation angle from t to
  g+ to g-
udsum = 0.0
ud2sum = 0.0
ph2=-120.0
do 1 i=1,3
ph2 = ph2+120.0
sph2=sin(ph2/57.29578)
cph2=cos(ph2/57.29578)
t2(2,1)=0.92718*cph2
t2(2,2)=-0.37461*cph2
t2(2,3)=sph2
t2(3,1)=0.92718*sph2
t2(3,2)=-0.37461*sph2
t2(3,3)=-cph2
→ Transformation of each backbone bond to
  reference frame of first bond and adding them
  to obtain distance between C1 and C6.
call mp(3,3,3,tt,t2,tt2)
call mp(3,3,1,tt2,lcc,ttl)
call add(3,h1,ttl,h2)
ph3=-120.0
do 2 j=1,3
ph3=ph3+120.0
sph3=sin(ph3/57.29578)
cph3=cos(ph3/57.29578)
t3(2,1)=0.92718*cph3
t3(2,2)=-0.37461*cph3
t3(2,3)=sph3
t3(3,1)=0.92718*sph3
t3(3,2)=-0.37461*sph3
t3(3,3)=-cph3
call mp(3,3,3,tt2,t3,tt23)
call mp(3,3,1,tt23,lcc,tttl)
call add(3,h2,tttl,h3)
ph4=-120.0
do 3 k=1,3
ph4=ph4+120.0
sph4=sin(ph4/57.29578)
cph4=cos(ph4/57.29578)
t4(2,1)=0.92718*cph4
```

```
      t4(2,2)=-0.37461*cph4
      t4(2,3)=sph4
      t4(3,1)=0.92718*sph4
      t4(3,2)=-0.37461*sph4
      t4(3,3)=-cph4
      call mp(3,3,3,tt23,t4,tt234)
      call mp(3,3,1,tt234,lcc,ttttl)
      call add(3,h3,ttttl,h4)
      → Distances between C1 and C6
      dc16=(h4(1,1)*h4(1,1)+h4(2,1)*h4(2,1)+h4(3,1) *h4
      (3,1))**0.5
      → Conformational probabilities
      v(i,j)= u2(i,j)
      call mp(3,3,3,u1,v,u4)
      v(i,j)=0.0
      v(j,k)= u3(j,k)
      call mp(3,3,3,u4,v,u5)
      prob=(u5(1,1)+u5(1,2)+u5(1,3))/z
      v(j,k)=0.0
      print 100,prob,dc16
      → Sum of Prob*Dist for each conformation
        to get average distance and distance²
        between C1 and C6
      udsum=udsum+prob*dc16
    3 ud2sum=ud2sum+prob*dc16*dc16
    2 continue
    1 continue
→ Output average C1-C6 distance and distance² and Z
      print 101,udsum,ud2sum,z
      stop
      end
      → Subroutine for multiplication of vectors
        a(r,s) and b(s,t) to give vector c(r,t)
      subroutine mp(r,s,t,a,b,c)
      integer r,s,t,ii,jj,kk
      real sum, a(r,s),b(s,t),c(r,t)
      do 5 ii=1,r
      do 5 jj=1,t
      sum=0
      do 7 kk=1,s
    7 sum=a(ii, kk)*b(kk,jj)+sum
    5 c(ii,jj)=sum
      return
      end
      → Subroutine for addition of vectors a
        and b to give vector c
```

```
      subroutine add(jj,a,b,c)
      integer i,jj
      real a(jj,1),b(jj,1),c(jj,1)
      do 1 i=1,jj
    1 c(i,1)=a(i,1)+b(i,1)
      return
      end
```

Results →

```
ttt
  0.573153 1.418585 0.000000 <-> (Tt)(1)
  1.529993 0.000000 0.000000 <-> (TtT2)(1)
  0.573151 1.418579 0.000000 <-> (TtT2T3)(1)
  1.529986 0.000000 0.000000 <-> (TtT2T3T4)(1)
  0.190290 6.399566 <-> probability d(1-6)

 ttg+
 -0.442931 0.797121 -1.228531 <-> (TtT2T3T4)(1)
 0.081832 5.374033 <-> probability d(1-6)

 ttg-
 -0.442931 0.797121 1.228531 <-> (TtT2T3T4)(1)
  0.081832 5.374033 <-> probability d(1-6)

 tg+t
 0.573151 -0.709289 0.000000 <-> (TtT2T3)(1)
 1.529986 0.000000 0.000000 <-> (TtT2T3T4)(1)
 0.081832 5.909089 <-> probability d(1-6)

 tg+g+
 -0.442931 0.665374 1.304596 <-> (TtT2T3T4)(1)
 0.035191 4.740186 <-> probability d(1-6)

 tg+g-
 -0.442931 -1.462494 0.076064 <-> (TtT2T3T4)(1)
 0.001204 4.053665 <-> probability d(1-6)

 tg-t
 0.573151 -0.709289 0.000000 <-> (TtT2T3)(1)
 1.529986 0.000000 0.000000 <-> (TtT2T3T4)(1)
 0.081832 5.909089 <-> probability d(1-6)

 tg-g+
 -0.442931 -1.462494 -0.076065 <-> (TtT2T3T4)(1)
 0.001204 4.053665 <-> probability d(1-6)

 tg-g-
 -0.442931 0.665374 -1.304595 <-> (TtT2T3T4)(1)
 0.035191 4.740187 <-> probability d(1-6)
```

```
g+tt
-0.442933 0.797124 -1.228531 <-> (TtT2)(1)
0.573151 1.418579 -1.228531 <-> (TtT2T3)(1)
-0.442931 0.797121 -1.228525 <-> (TtT2T3T4)(1)
0.081832 5.374030 <-> probability d(1-6)

g+tg+
-0.442935 0.797122 1.228528 <-> (TtT2T3T4)(1)
0.035191 4.779442 <-> probability d(1-6)

g+tg-
 1.529990 -0.000002 -0.000003 <-> (TtT2T3T4)(1)
 0.035191 5.374032 <-> probability d(1-6)

g+g+t
0.450997 -0.659935 -1.228531 <-> (TtT2T3)(1)
-0.442931 0.797121 -1.228525 <-> (TtT2T3T4)(1)
0.035191 4.740181 <-> probability d(1-6)

g+g+g+
1.522427 -0.128693 -0.080777 <-> (TtT2T3T4)(1)
0.015134 4.698130 <-> probability d(1-6)

g+g+g-
-0.572653 -1.410083 -0.156841 <-> (TtT2T3T4)(1)
0.000518 3.102311 <-> probability d(1-6)

g+g-t
-1.521929 0.137189 -1.228531 <-> (TtT2T3)(1)
-0.442931 0.797121 -1.228525 <-> (TtT2T3T4)(1)
0.001204 4.053657 <-> probability d(1-6)

g+g-g+
-0.694806 -1.360729 -0.080777 -1.228525 <-> (TtT2T3T4)(1)
0.000018 1.792586 <-> probability d(1-6)

g+g-g-
-0.5-2652 0.717785 1.223819 <-> (TtT2T3T4)(1)
0.000518 3.102305 <-> probability d(1-6)

g-tt
-0.442933 0.797124 1.228531 <-> (TtT2)(1)
0.573151 1.418579 1.228531 <-> (TtT2T3)(1)
-0.442931 0.797121 1.228525 <-> (TtT2T3T4)(1)
0.081832 5.374029 <-> probability d(1-6)
```

```
g-tg+
1.529990 -0.000002 0.000003 <-> (TtT2T3T4)(1)
0.035191 5.374032 <-> probability d(1-6)

g-tg-
-0.442936 0.797123 -1.228528 <-> (TtT2T3T4)(1)
0.035191 4.779442 <-> probability d(1-6)

g-g+t
-1.521929 0.137189 1.228531 <-> (TtT2T3)(1)
-0.442931 0.797121 1.228526 <-> (TtT2T3T4)(1)
0.001204 4.053657 <-> probability d(1-6)

g-g+g+
-0.572653 0.717785 -1.223819 <-> (TtT2T3T4)(1)
0.000518 3.102304 <-> probability d(1-6)

g-g+g-
-0.694806 -1.360729 0.080777 <-> (TtT2T3T4)(1)
0.000018 1.792587 <-> probability d(1-6)

g-g-t
0.450997 -0.659935 1.228531 <-> (TtT2T3)(1)
-0.442931 0.797121 1.228525 <-> (TtT2T3T4)(1)
0.035191 4.740181 <-> probability d(1-6)

g-g-g+
-0.572652 -1.410083 0.156841 <-> (TtT2T3T4)(1)
0.000518 3.102311 <-> probability d(1-6)

g-g-g-
1.522427 -0.128693 0.080777 <-> (TtT2T3T4)(1)
0.015134 4.698130 <-> probability d(1-6)


{5.494033 30.531794 5.255144 <-> <d(1-6> <[d(1-6)²]> Z}
```

Fortran Program for Hexane and PE Matrix Method Calculations of Average dimensions $<r^2>_o$.

```
→ Declaration of variables
real l(3,1),lt(1,3),l2,thet, phi, sphi, cphi,
gp(15,15),
7t1(3,3),t2(3,3),t3(3,3),tbars(9,9),e3(3,3), jf(1,15),
```

```
7jl(15,1),g(15,15),u(3,3),l2u(3,3),ult(3,9),
ubt(3,9),
7uxl(9,3),cr, sthet, cthet, z,ue3(9,9),uet(9,9),
up(3,3),
7jfg(1,15),r2,u1(3,3),u2(3,3),g1(15,15),t,rt,s,w,
7g2(15,15),u4(3,3),u32(3,3),u64(3,3),u96(3,3),
u100(3,3),
7g4(15,15),g32(15,15),g64(15,15),g96(15,15),
g100(15,15),
7u12(3,3),u123(3,3),g12(15,15),g123(15,15),
u1234(3,3),
7u12345(3,3),g1234(15,15),g12345(15,15),upp(3,3),
7gpp(15,15)
integer i, j
→ Output formats
101 format(f10.3)
102 format(e8.3)
→ T, RT, statistical weights, U matrices
t=298.16
rt=1.9872*t
s=1.0/(2.71828**(500.0/rt))
w=1.0/(2.71828**(2000.0/rt))
u(1,1)=1.0
u(1,2)=s
u(1,3)=s
u(2,1)=1.0
u(2,2)=s
u(2,3)=s*w
u(3,1)=1.0
u(3,2)=s*w
u(3,3)=s
up(1,1)=1.0
up(1,2)=s
up(1,3)=s
up(2,1)=1.0
up(2,2)=s
up(2,3)=s
up(3,1)=1.0
up(3,2)=s
up(3,3)=s
upp(1,1)=1.0
upp(1,2)=1.0
upp(1,3)=1.0
upp(2,1)=1.0
upp(2,2)=1.0
upp(2,3)=1.0
upp(3,1)=1.0
```

```
upp(3,2)=1.0
upp(3,3)=1.0
→ Column and row bond vectors
l(1,1)=1.54
l(2,1)=0.0
l(3,1)=0.0
lt(1,1)=1.54
lt(1,2)=0.0
lt(1,3)=0.0
→ Transformation matrices for t, g+, g- conformations
thet=68.0
sthet=sin(thet/57.29578)
cthet=cos(thet/57.29578)
phi=0.0
sphi=sin(phi/57.29578)
cphi=cos(phi/57.29578)
t1(1,1)=cthet
t1(1,2)=sthet
t1(1,3)=0.0
t1(2,1)=sthet*cphi
t1(2,2)=-cthet*cphi
t1(2,3)=sphi
t1(3,1)=sthet*sphi
t1(3,2)=-cthet*sphi
t1(3,3)=-cphi
phi=120.0
sphi=sin(phi/57.29578)
cphi=cos(phi/57.29578)
t2(1,1)=cthet
t2(1,2)=sthet
t2(1,3)=0.0
t2(2,1)=sthet*cphi
t2(2,2)=-cthet*cphi
t2(2,3)=sphi
t2(3,1)=sthet*sphi
t2(3,2)=-cthet*sphi
t2(3,3)=-cphi
phi=240.0
sphi=sin(phi/57.29578)
cphi=cos(phi/57.29578)
t3(1,1)=cthet
t3(1,2)=sthet
t3(1,3)=0.0
t3(2,1)=sthet*cphi
t3(2,2)=-cthet*cphi
t3(2,3)=sphi
t3(3,1)=sthet*sphi
```

```
t3(3,2)=-cthet*sphi
t3(3,3)=-cphi
→ Constructing the 15 × 15 G matrices
call foel(1,9,1,9,tbars,0.0,9,9)
call fill(1,1,3,3,t1,tbars,3,3,9,9)
call fill(4,4,3,3,t2,tbars,3,3,9,9)
call fill(7,7,3,3,t3,tbars,3,3,9,9)
call foel(1,3,1,3,e3,0.0,3,3)
e3(1,1)=1.0
e3(2,2)=1.0
e3(3,3)=1.0
call foel(1,1,1,15,jf,0.0,1,15)
jf(1,1)=1.0
call foel(1,15,1,1,jl,0.0,15,1)
jl(13,1)=1.0
jl(14,1)=1.0
jl(15,1)=1.0
call foel(1,15,1,15,g,0.0,15,15)
call foel(1,15,1,15,gp,0.0,15,15)
call foel(1,15,1,15,gpp,0.0,15,15)
l2=(l(1,1)*l(1,1))/2.0
do 1 i=1,3
do 1 j=1,3
1 l2u(i,j)=l2*u(i, j)
call fill(1,1,3,3,u,g,3,3,15,15)
call fill(13,13,3,3,u,g,3,3,15,15)
call fill(1,13,3,3,l2u,g,3,3,15,15)
call dp(3,3,1,3,u,lt,ult,3,9)
call mp(3,9,9,ult,tbars,ubt)
call fill(1,4,3,9,ubt,g,3,9,15,15)
call dp(3,3,3,3,u,e3,ue3,9,9)
call mp(9,9,9,ue3,tbars,uet)
call fill(4,4,9,9,uet,g,9,9,15,15)
call dp(3,3,3,1,u,l,uxl,9,3)
call fill(4,13,9,3,uxl,g,9,3,15,15)
do 2 i=1,3
do 2 j=1,3
2 l2u(i,j)=l2*up(i,j)
call fill(1,1,3,3,up,gp,3,3,15,15)
call fill(13,13,3,3,up,gp,3,3,15,15)
call fill(1,13,3,3,l2u,gp,3,3,15,15)
call dp(3,3,1,3,up,lt,ult,3,9)
call mp(3,9,9,ult,tbars,ubt)
call fill(1,4,3,9,ubt,gp,3,9,15,15)
call dp(3,3,3,3,up,e3,ue3,9,9)
call mp(9,9,9,ue3,tbars,uet)
call fill(4,4,9,9,uet,gp,9,9,15,15)
```

```
  call dp(3,3,3,1,up,l,uxl,9,3)
  call fill(4,13,9,3,uxl,gp,9,3,15,15)
  do 3 i=1,3
  do 3 j=1,3
3 l2u(i,j)=l2*upp(i,j)
  call fill(1,1,3,3,upp,gpp,3,3,15,15)
  call fill(13,13,3,3,upp,gpp,3,3,15,15)
  call fill(1,13,3,3,l2u,gpp,3,3,15,15)
  call dp(3,3,1,3,upp,lt,ult,3,9)
  call mp(3,9,9,ult,tbars,ubt)
  call fill(1,4,3,9,ubt,gpp,3,9,15,15)
  call dp(3,3,3,3,upp,e3,ue3,9,9)
  call mp(9,9,9,ue3,tbars,uet)
  call fill(4,4,9,9,uet,gpp,9,9,15,15)
  call dp(3,3,3,1,upp,l,uxl,9,3)
  call fill(4,13,9,3,uxl,gpp,9,3,15,15)
  → Multiplying U matrices
  call mp(3,3,3,upp,up,u12)
  call mp(3,3,3,u12,u,u123)
  call mp(3,3,3,u123,u,u1234)
  call mp(3,3,3,u1234,upp,u12345)
  → Partition function
  z=u12345(1,1)+u12345(1,2)+u12345(1,3)
  print 102, z
  → Multiplication of G Matrices
  call mp(15,15,15,gpp,gp,g12)
  call mp(15,15,15,g12,g,g123)
  call mp(15,15,15,g123,g,g1234)
  call mp(15,15,15,g1234,gpp,g12345)
  call mp(1,15,15,jf,g12345,jfg)
  call mp(1,15,1,jfg,jl,r2)
  → Characteristic ratio = <r²>o /nl²
  cr=(2.0*r2)/(5.0*1.54*1.54*z)
  print 101, cr
  → Z for 100 bond PE
  call mp(3,3,3,u,u,u1)
  call mp(3,3,3,u1,u1,u4)
  call mp(3,3,3,u4,u4,u1)
  call mp(3,3,3,u1,u1,u2)
  call mp(3,3,3,u2,u2,u32)
  call mp(3,3,3,u32,u32,u64)
  call mp(3,3,3,u64,u32,u96)
  call mp(3,3,3,u96,u4,u100)
  z=u100(1,1)+u100(1,2)+u100(1,3)
  print 102, z
  → Product of 100 G matrices
  call mp(15,15,15,g,g,g1)
```

```
call mp(15,15,15,g1,g1,g4)
call mp(15,15,15,g4,g4,g1)
call mp(15,15,15,g1,g1,g2)
call mp(15,15,15,g2,g2,g32)
call mp(15,15,15,g32,g32,g64)
call mp(15,15,15,g64,g32,g96)
call mp(15,15,15,g96,g4,g100)
call mp(1,15,15,jf,g100,jfg)
call mp(1,15,1,jfg,jl,r2)
→ <r²>o =/nl² for 100 bond PE
cr=(2.0*r2)/(100.0*1.54*1.54*z)
print 101, cr
stop
end
→ Subroutine that fills a matrix with all elements = val
subroutine foel(a,b,c,d,aa,val,d1,d2)
integer a,b,c,d,ii,jj,d1,d2
real val,aa(d1,d2)
do 8 ii=a,b
do 8 jj=c,d
8 aa(ii,jj)=val
return
end
→ Subroutine to fill a matrix aaa (d21,d22) with a
  matrix aa(d11,d12) beginning at the (a,b) element
  of aaa.
subroutine fill(a,b,c,d,aa,aaa,d11,d12,d21,d22)
integer ii,jj,a,b,c,d,d11,d12,d21,d22,e,f
real aa(d11,d12),aaa(d21,d22)
do 4 ii=1,c
do 4 jj=1,d
e=a-1+ii
f=b-1+jj
4 aaa(e,f)=aa(ii,jj)
return
end
→ Subroutine that multiplies matrices a(r,s)xb(s,t)
  = c(r,t)
subroutine mp(r,s,t,a,b,c)
integer r,s,t,ii,jj,kk
real sum, a(r,s),b(s,t),c(r,t)
do 5 ii=1,r
do 5 jj=1,t
sum=0
do 7 kk=1,s
7 sum=a(ii,kk)*b(kk,jj)+sum
5 c(ii, jj)=sum
```

```
return
end
→ Subroutine that takes the direct product of
  aa(m,n) and bb(p,s) to give cc(mxp,nxs)
subroutine dp(m,n,p,s,aa,bb,cc,pr1,pr2)
integer m,n,p,s,ii,jj,kk,ll,pr1,pr2,a,b
real aa(m,n),bb(p,s),cc(pr1,pr2)
do 2 ii=1,m
do 2 jj=1,n
do 2 kk=1,p
do 2 ll=1,s
a=((ii-1)*p)+kk
b=((jj-1)*s)+ll
2 cc(a, b)=aa(ii,jj)*bb(kk,ll)
return
end
```

Conversion of the above Fortran programs into more modern code should be relatively easy. For example, Matlab has the function for conducting the requisite matrix mathematics used in the Fortran program.

4 Experimental Determination of Polymer Microstructures with ^{13}C-NMR Spectroscopy

INTRODUCTION

Chemists and other material scientists and engineers are interested in establishing the connections between the structures of molecules and the properties of materials made from them, which is the overarching theme of this book. Almost 30 years ago I wrote another book, *NMR Spectroscopy and Polymer Microstructure*, which was subtitled *The Conformational Connection*, for the purpose of demonstrating how the ^{13}C-NMR spectra of polymers can be assigned to their contributing microstructures (Tonelli 1989). Then and now, ^{13}C-NMR was and remains (Tonelli 2017) the most sensitive probe of molecular structure, including the microstructures of polymers. To fully utilize ^{13}C-NMR, the resonance frequencies of peaks appearing in the resulting spectra must be assigned to the molecular structures that generated them.

However, even the most advanced quantum mechanical methods cannot yet estimate ^{13}C-NMR resonance frequencies accurately enough to delineate the detailed molecular structures that produce them (Tonelli 2017). The arrangement of proximal electrons and the local magnetic fields they produce by their movements shield NMR-active ^{13}C nuclei from the applied static magnetic field, thereby affecting their resonance frequencies. Unfortunately, even the most sophisticated quantum mechanical methods cannot yet describe with sufficient accuracy the structural dependence of the electronic shielding of magnetic nuclei. Particularly for flexible molecules like polymers, this is the case, because the magnetic shielding of nuclei belonging to particular polymer microstructures must not only be accurately predicted, but also must be accurately calculated for each and averaged over all myriad conformations that are generated by each microstructure.

SUBSTITUENTS EFFECTS

Instead, empirical nuclear shielding effects were deduced from the ^{13}C-NMR spectra of model compounds with known structures (Stothers 1972; Bovey 1974). Examples of the shielding of aliphatic ^{13}C nuclei produced by protonated carbons that are α, β, and γ substituents are presented in Tables 4.1–4.3, respectively. These nuclear shielding substituent effects were then and remain successfully used to make connections between polymer microstructures and their ^{13}C-NMR spectra.

As an example of their use, we can evaluate the relative ^{13}C resonances expected in polypropylene (PP) by counting the numbers of carbon substituents that are α, β, and γ to the CH; CH_2; and CH_3 carbons in the **bold** repeat unit below, which are, respectively, 3, 2, and 4; 2, 4, and 2; and 1, 2, and 2. We note from Tables 4.1–4.3 that average α-, β-, and γ-effects are, respectively, ~+9, +9, and −2 ppm. Thus, the relative resonance frequencies of CH; CH_2; and CH_3 carbons estimated using these substituent effects are 3(9) + 2(9) + 4(−2) = 37 ppm; 2(9) + 4(9) + 2(−2) = 50 ppm; and 1(9) + 2(9) + 2(−2) = 23 ppm. As a result, we expect the CH_3 carbon to resonate most upfield (23 ppm), while the CH and CH_2 carbons should come, respectively, downfield from it by ~14 and ~27 ppm. In the spectra in Figure 4.1, we see that this is in fact the order of ^{13}C resonances in atactic-PP.

[PP=(—CH$_2$—CH(CH$_3$)—CH$_2$—CH(CH$_3$)—**CH$_2$—CH(CH$_3$)**—CH$_2$—CH(CH$_3$)—CH$_2$—)]

Principal in importance among these substituent effects is the nuclear shielding produced by γ-substituents, which has been demonstrated to have a conformational origin (Tonelli 1989). A γ-substituent was found to shield a ^{13}C nucleus if the central bond between them produced a proximal arrangement by adopting a *gauche* conformation (see Figure 4.2). In Figure 4.3, the methyl carbon resonance frequencies observed in butane, 1-propanol, and 1-chloropropane are compared, along with the *gauche* population of their central bonds. On the assumption that a *trans* arrangement produces no nuclear shielding, in each case, they enable an estimate of the nuclear shielding produced by their γ-substituents (CH_3,OH,Cl) when arranged proximally in a *gauche* conformation.

The ^{13}C-NMR spectra observed for PPs with different stereosequences in Figure 4.1 clearly show that the resonance frequencies for all three carbon types do show a sensitivity to their stereosequences. Different resonance frequencies are observed for stereoregular i- and s-PPs, and all three carbon types in stereoirregular a-PP evidence multiple resonances. However, independent of stereosequence, the three carbon types have the same numbers and types of α-, β-, and γ-substituents. This suggests that

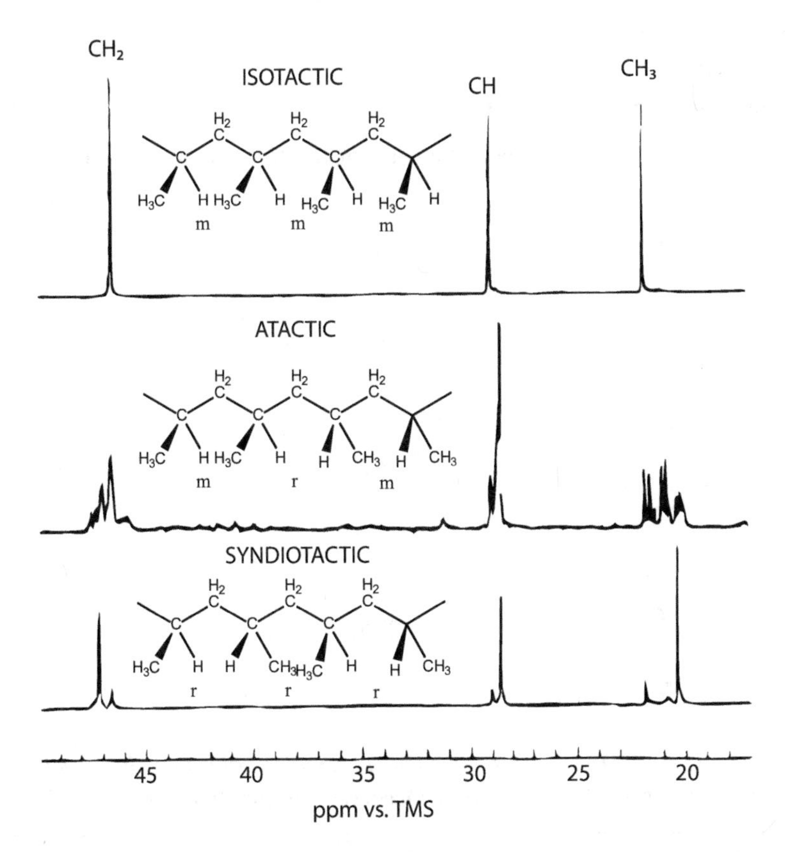

FIGURE 4.1 ^{13}C-NMR spectra observed at 25 MHz of three PP samples with different stereosequences or tacticities [Isotactic (all *m* diads), syndiotactic (all *r* diads), and atactic (random distribution of *m* and *r* diads)]. (Reprinted with permission from Tonelli, A. E. and Schilling, F. C., *Acc. Chem. Res.*, 14, 233, 1981. Copyright 1981 American Chemical Society.)

their stereosequences affect the amounts of γ-*gauche* shielding experienced by their ^{13}C nuclei, and this can only be the case if their conformations are sensitive to their stereosequences or tacticities. That this is the case is shown in Table 4.4. There, the probability of finding the 5th backbone bond in the *trans* conformation in each of the various PP pentad stereosequences is presented. The likelihood of finding this bond in the *trans* conformation changes by 80% as the PP pentad stereosequence is altered (Suter and Flory 1975; Schilling and Tonelli 1980; Tonelli and Schilling 1981). As a consequence, significant differences in γ-*gauche* shielding and ^{13}C resonant frequencies are expected for the ^{13}C nuclei in different a-PP stereosequences.

When the stereosequence-dependent γ-*gauche* shieldings of methyl carbons in atactic PP are estimated as indicated in Figure 4.4, using the rotational isomeric state (RIS) model developed by Suter and Flory, we

TABLE 4.1
α-Substituent Effect on δ¹³C

		δ¹³C from Tetramethylsilane (TMS) (ppm)	α-effect (ppm)
(a)	$^{\circ}CH_3—H$	–2.1	—
(b)	$^{\circ}CH_3—{}^{\alpha}CH_3$	5.9	8.0
(c)	$^{\circ}CH_2({}^{\alpha}CH_3)_2$	16.1	10.2
(d)	$^{\circ}CH({}^{\alpha}CH_3)_3$	25.2	9.1
(e)	$^{\circ}C({}^{\alpha}CH_3)_4$	27.9	2.7

Source: Bovey, F. A., *Proceedings of the International Symposium on Macro-molecules*, Rio de Janeiro, July 26–31, E. B. Mano, Ed., Elsevier, New York, 169, 1974.

TABLE 4.2
β-Substituent Effect on δ¹³C

		δ¹³C from TMS (ppm)	β-Effect (ppm)
(a)	$^{\circ}CH_3—{}^{\alpha}CH_3$	5.9	—
(b)	$^{\circ}CH_3—{}^{\alpha}CH_2—{}^{\beta}CH_3$	15.6	9.7
(c)	$^{\circ}CH_3—{}^{\alpha}CH({}^{\beta}CH_3)_2$	24.3	8.7
(d)	$^{\circ}CH_3—{}^{\alpha}C({}^{\beta}CH_3)_3$	31.5	7.2

Source: Bovey, F. A., *Proceedings of the International Symposium on Macro-molecules*, Rio de Janeiro, July 26–31, E. B. Mano, Ed., Elsevier, New York, 169, 1974.

TABLE 4.3
γ-Substituent Effect on δ¹³C

		δ¹³C from TMS (ppm)	γ-effect (ppm)
(a)	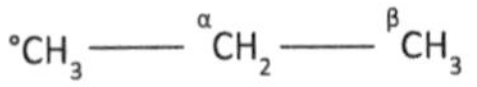 $^{\circ}CH_3-{}^{\alpha}CH_2-{}^{\beta}CH_3$	15.6	—
(b)	 $^{\circ}CH_3-{}^{\alpha}CH_2-{}^{\beta}CH-{}^{\gamma}CH_3$	13.2	−2.4
(c)	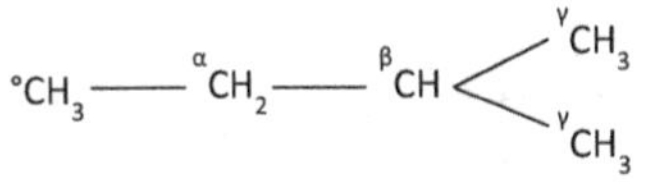 $^{\circ}CH_3-{}^{\alpha}CH_2-{}^{\beta}CH({}^{\gamma}CH_3)_2$	11.5	−1.7
(d)	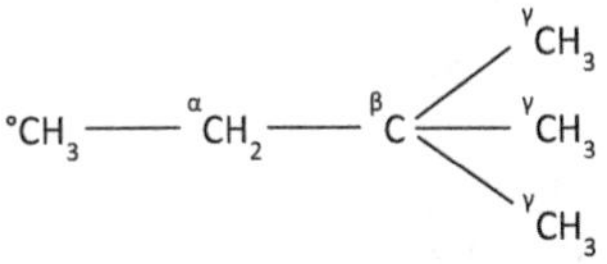 $^{\circ}CH_3-{}^{\alpha}CH_2-{}^{\beta}C({}^{\gamma}CH_3)_3$	8.7	−2.8
(e)	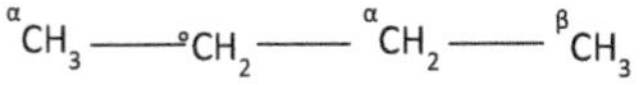 $^{\alpha}CH_3-{}^{\circ}CH_2-{}^{\alpha}CH_2-{}^{\beta}CH_3$	25.0	
(f)	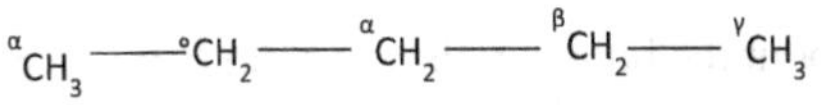 $^{\alpha}CH_3-{}^{\circ}CH_2-{}^{\alpha}CH_2-{}^{\beta}CH_2-{}^{\gamma}CH_3$	22.6	−2.4
(g)	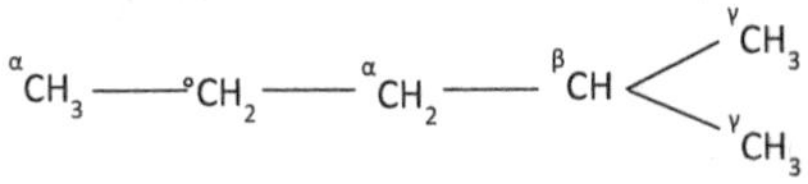 $^{\alpha}CH_3-{}^{\circ}CH_2-{}^{\alpha}CH_2-{}^{\beta}CH({}^{\gamma}CH_3)_2$	20.7	−1.9
(h)	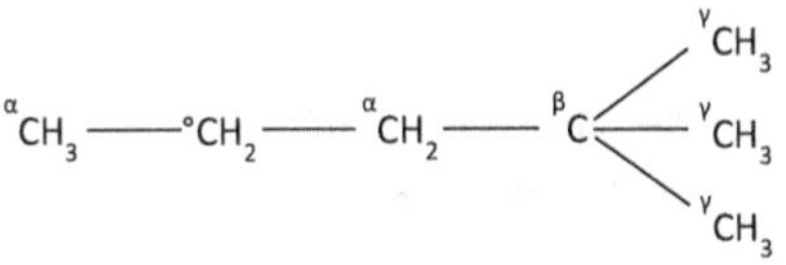 $^{\alpha}CH_3-{}^{\circ}CH_2-{}^{\alpha}CH_2-{}^{\beta}C({}^{\gamma}CH_3)_3$	18.8	−1.9

Source: Bovey, F. A., *Proceedings of the International Symposium on Macro-molecules*, Rio de Janeiro, July 26–31, E. B. Mano, Ed., Elsevier, New York, 169, 1974.

obtain the predicted methyl carbon region of the ¹³C-NMR spectrum seen in Figure 4.5b, which is in close agreement with the observed spectrum above in Figure 4.5a (Suter and Flory 1975; Shilling and Tonelli 1980; Tonelli and Schilling 1981).

Additional evidence that the stereosequence-dependent ¹³C-NMR resonance frequencies are produced by stereosequence-dependent conformations as manifested by their γ-*gauche* shieldings is offered by propylene-vinyl chloride (P-VC) copolymers with low P content (Tonelli and Schilling 1984). An example would be the P-VC pentad with an isolated propylene unit.

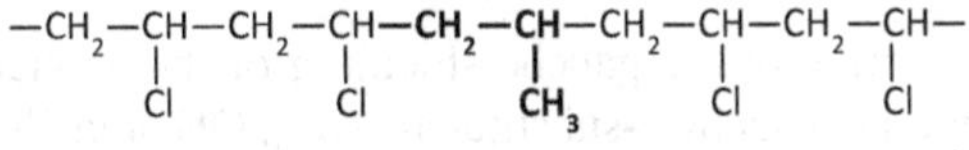

$$-CH_2-\underset{Cl}{CH}-CH_2-\underset{Cl}{CH}-\mathbf{CH_2}-\underset{\mathbf{CH_3}}{\mathbf{CH}}-CH_2-\underset{Cl}{CH}-CH_2-\underset{Cl}{CH}-$$

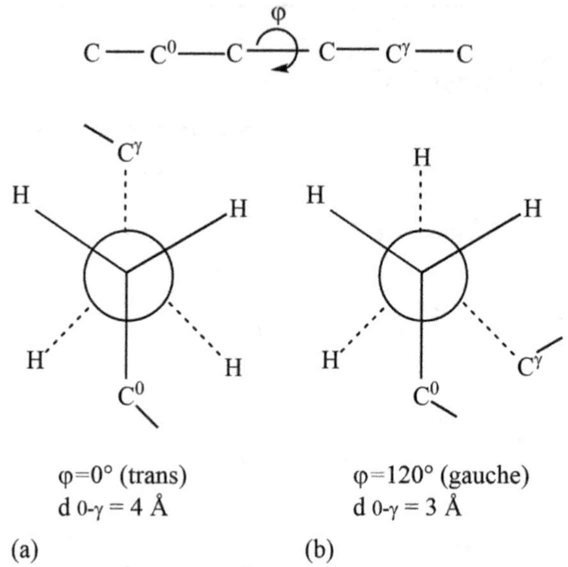

$C—C^0—C—C—C^\gamma—C$ (rotation φ)

$\varphi = 0°$ (trans)
$d_{0\text{-}\gamma} = 4$ Å

(a)

$\varphi = 120°$ (gauche)
$d_{0\text{-}\gamma} = 3$ Å

(b)

FIGURE 4.2 Newman projections about the bonds in PE in the (a) *trans* ($\varphi = 0°$) and (b) gauche ($\varphi = 120°$) conformations and the distances between carbons γ to each other.

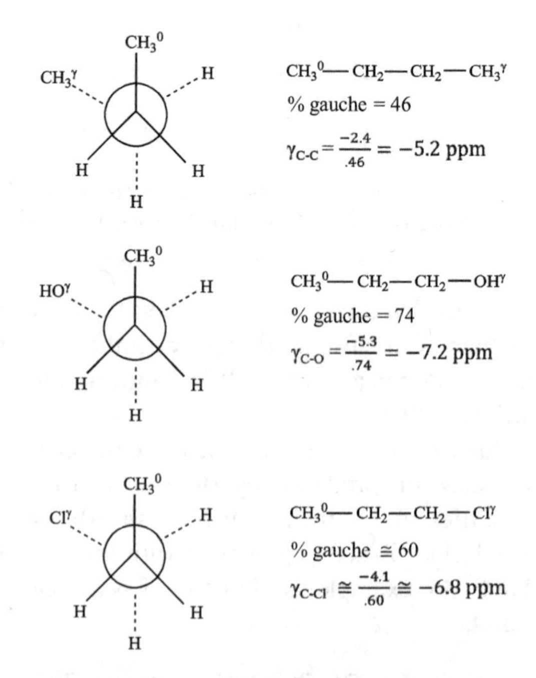

FIGURE 4.3 Derivation of the gauche shielding of the applied magnetic field produced at a carbon nucleus by γ-substituents CH_3, OH, and Cl.

TABLE 4.4
Calculated Probabilities that the 5th Bond in PP Pentad Stereosequences is *trans*

```
 C m,r C m,r C   m,r  C m,r C
 |     |     |        |     |
—C—C—C—C—C—C—C—φ—C—C—C—C—C—
```

Pentad Stereosequence	$P(\varphi = t)$
mrmr	0.440
rrmr	0.472
mmmm	0.523
rmmr	0.539
rmmm	0.582
rrrr	0.635
mrrm	0.685
rrrm	0.712
mmrr	0.742
rmrm	0.763
mmrm	0.792

Source: Suter, U. W. and Flory, P. T., *Macromolecules*, 8, 765, 1975.

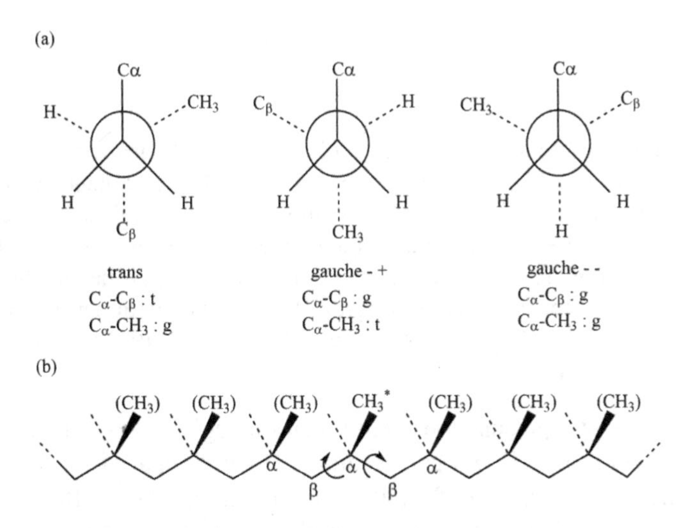

FIGURE 4.4 (a) Newman diagrams of a four carbon fragment of PP in different $-CH_2-C_\alpha-$ bond conformations illustrating γ-*gauche* arrangements of their carbon nuclei. (b) PP heptad indicating backbone rotations α_S which alter the *gauche* arrangements of the central CH_3* and its γ-substituents, the nearest neighbor C_αs [see (a)] and their nuclear shielding].

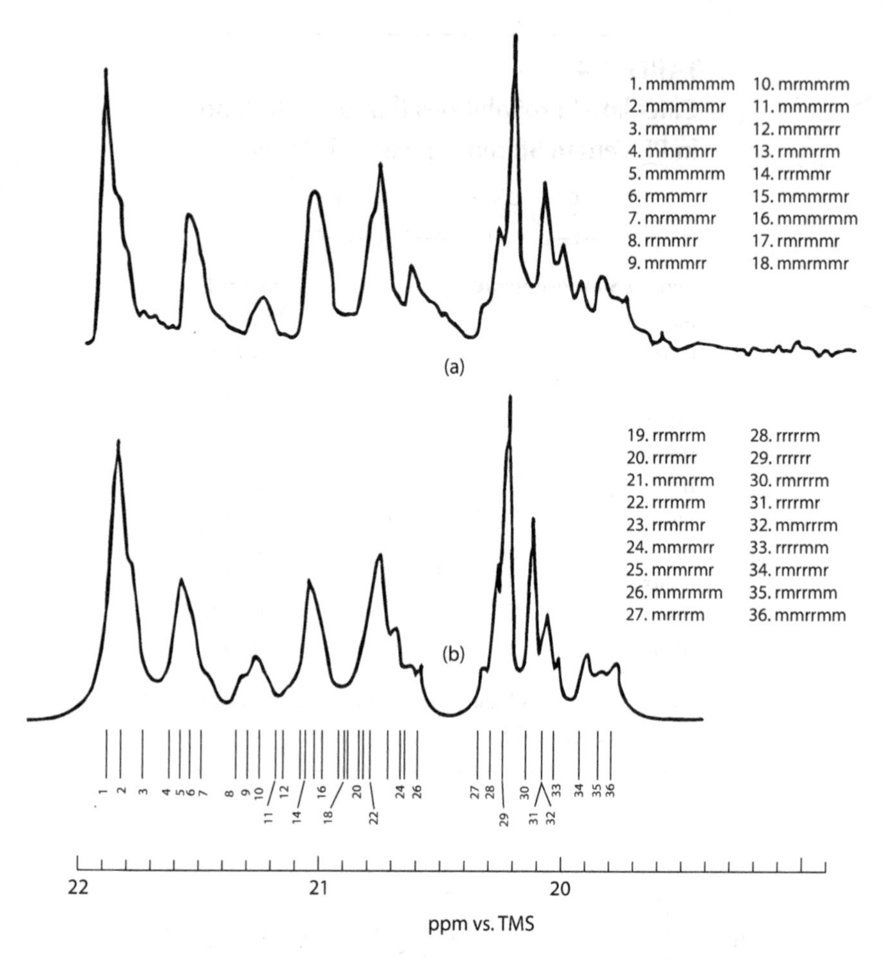

FIGURE 4.5 (a) Methyl carbon region of the ^{13}C-NMR spectrum of the same atactic PP shown in Figure 4.1, but recorded at 90 MHz in *n*-heptane at 67°C. (b) Simulated methyl carbon spectrum obtained from chemical shifts calculated using the γ-gauche-effect method, as represented by the line spectrum below, and assuming Lorentzian peaks of <0.1 ppm width at half-height (Schilling and Tonelli 1980; Tonelli and Schilling 1981). (Reprinted with permission from Tonelli, A. E. and Schilling, F. C., *Acc. Chem. Res.*, 14, 233, 1981. Copyright 1981 American Chemical Society.)

In Figure 4.6, we present the methyl carbon region of an atactic P-VC copolymer with 5 mol% propylene units. When compared to the methyl carbon region of the atactic PP spectrum in Figure 4.5, two differences are evident. First, the isolated propylene methyl carbons in atactic P-VC only show sensitivity to pentad stereosequences, while in PP, the resonances of the methyl carbons are sensitive to heptad stereosequences. The source of this difference is demonstrated in Table 4.5, where relative ^{13}C-NMR resonance frequencies calculated for the methyl carbons in several heptad

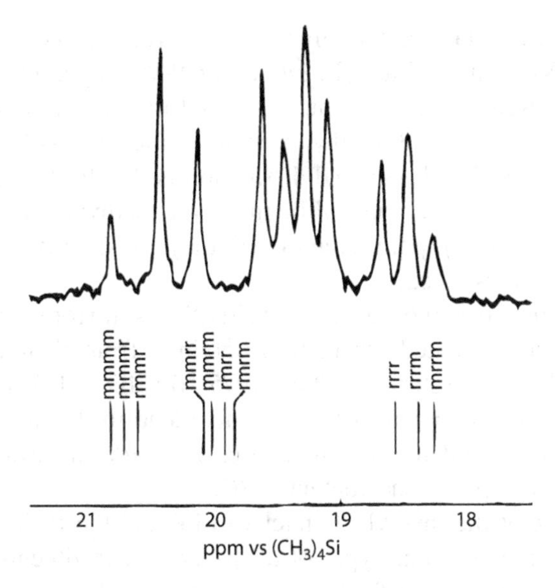

FIGURE 4.6 The methyl carbon region of the 50 MHz ^{13}C-NMR spectrum of an isolated propylene unit in a low-P content P-VC copolymer. (Reprinted with permission from Tonelli, A. E. and Schilling, F. C., *Macromolecules*, 17, 1946, 1984. Copyright 1984 American Chemical Society.)

TABLE 4.5
Relative ^{13}C-NMR Resonance Frequencies Calculated for the P-VC and PP Methyl Carbons in Several Heptad Stereosequences Containing the Same Pentad Stereosequence

	Δδ[a], ppm	
Heptad	**P-VC**	**PP**
r(rmrm)r	0	0
m(rmrm)r	−0.01	−0.07
r(rmrm)m	−0.01	−0.05
m(rmrm)m	−0.03	−0.10
r(mrrm)r	0	0
m(mrrm)r	−0.04	−0.07
m(mrrm)m	−0.07	−0.12

Source: Tonelli, A. E. and Schilling, F. C., *Macromolecules*, 17, 1946, 1984.

[a] Δδ is the difference in chemical shift among the various heptads containing the same central pentad stereosequence. $\gamma_{CH3,CH} = -5$ ppm was used for both PP and P-VC.

stereosequences containing the same pentad stereosequence are compared for atactic P-VC and PP. The RIS model for P-VC copolymers developed by Mark was used to calculate the bond conformational populations and γ-*gauche* shielding of the P methyl carbons in the various P-VC stereosequences (Mark 1973). The second distinction is that P-VC methyl carbon resonances range over 2.5 ppm, while in PP, the methyl carbon resonances are spread over only 2.0 ppm, even though they are sensitive to the longer-range heptad stereosequences.

The difference in stereosequence sensitivities is mirrored in the relative γ-*gauche*-effect calculated resonances obtained from their distinct RIS models (Mark 1973; Suter and Flory 1975; Tonelli and Schilling 1984). These same conformational differences also lead to the larger spread in stereosequence-dependent resonances observed for the isolated methyl carbons in the low propylene content P-VC.

Remember that the methyl ^{13}C nuclei in PP and the P-VC copolymers have the same numbers and types of α-, β-, and γ-substituents, so the differences in their resonance frequencies must depend solely on their conformational differences leading to distinct amounts of γ-*gauche*-effect shielding.

Through these few examples, we have tried to demonstrate that the local microstructure of a polymer affects its local conformation, the local conformation affects the shielding experienced by ^{13}C nuclei that are located in the local polymer microstructure, and, consequently, affect their resonance frequencies.

$$\textit{Microstructure} \rightarrow \textit{Conformation} \rightarrow B_i = B_o(1 - \sigma) \rightarrow \delta^{13}C_i$$

The γ-gauche shielding of ^{13}C nuclei enables us to both complete and simplify the conformational connection between the microstructures and ^{13}C-NMR spectra of polymers leading to the following:

Golden Rule →

```
                    γ-gauche effect
Microstructures ----------------> δ13Cs.
```

Though we have demonstrated how ^{13}C-NMR, currently the method of choice, can be used to determine polymer microstructures, we must not fail to point out what important and potentially crucial structural information this and all other presently used characterization methods fail to provide. While they all can, to one degree or another, tell us what and how much of each local short-range constituent microstructure are present in polymer samples, they cannot locate the microstructures along or between chains. Any means of polymer structural characterization that has a limited short-range sensitivity cannot tell where along a polymer chain its microstructural elements reside. Are they distributed uniformly or segregated along the polymer chains, and do all the chains in a polymer sample have similar average structures, or is the sample structurally heterogeneous?

In the Appendix, we describe an experimental means that begins to answer these critical structural questions. That means, the birefringence produced by a polymer solute in a dilute solution that is subject to a strong electric field, was first discovered by John Kerr and later called the Kerr effect. We need only remember that it is the primary structures of proteins which determine their biological functions, and not just the overall compositions of their amino acids, to justify a search for a means to also determine the complete molecular architectures or macrostructures of synthetic polymers.

REFERENCES

Bovey, F. A. (1974), *Proceedings of the International Symposium on Macromolecules*, Rio de Janeiro, July 26–31, E. B. Mano, Ed., Elsevier, New York, p. 169.

Mark, J. E. (1973), *J. Polym. Sc. Polym. Phys. Ed.*, 11, 1375.

Schilling, F. C., Tonelli, A. E. (1980), *Macromolecules*, 13, 270.

Stothers, J. B. (1972), *Carbon-13 NMR Spectroscopy*, Academic Press, London, UK.

Suter, U. W., Flory, P. T. (1975), *Macromolecules*, 8, 765.

Tonelli, A. E., Schilling, F. C. (1981), *Acc. Chem. Res.*, 14, 233.

Tonelli, A. E., Schilling, F. C. (1984), *Macromolecules*, 17, 1946.

Tonelli, A. E. (1989), *NMR Spectroscopy and Polymer Microstructure: The Conformational Connection*, Wiley-VCH, New York.

Tonelli, A. E. (2017), "From NMR Spectra to Molecular Structures and Conformations" in *Stereochemistry and Global Connectivity: The Legacy of Ernest Eliel*, ACS Symposium Series # XYZ, American Chemical Society, Washington, DC.

DISCUSSION QUESTIONS

1. Why is determining the microstructures that are present in polymer chains necessary to begin understanding the behaviors of their materials?
2. Currently, which experimental probe is most sensitive to polymer microstructures, and why is this the case?
3. Is there an *a priori* means for predicting the observable outcomes for distinct polymer microstructures obtained from the experimental probe in question 2? If there is, what is that means, and if not, why not?
4. Describe how the conformational consequences of different polymer microstructures are reflected by the experimental probe in question 2 and how they can be identified and assigned to each microstructure?

5. How might one use the experimental probe in question 2 to distinguish between a blend of polypropylene and poly(1-butene) (PB) and a PP/PB copolymer?

$$[PP = -(-CH_2\text{-}\underset{\displaystyle CH_3}{\underset{|}{CH}}\text{-})_n\text{-}] \quad [PB = -(-CH_2\text{-}\underset{\displaystyle CH_2\text{-}CH_3}{\underset{|}{CH}}\text{-})_n\text{-}]$$

6. Though the experimental probe asked for in question 2 is currently the most sensitive to polymer microstructures, what important structural features of polymers is it unable to distinguish and identify, and why?

APPENDIX 4.1: POLYMER MACROSTRUCTURES AND THE KERR EFFECT

While spectroscopic probes sensitive only to local polymer structures, like NMR, can identify and quantify short-range microstructural elements, they are unable to locate their positions along the polymer backbone. The complete molecular architectures of synthetic polymers, which may be called their macrostructures, consist of the types and amounts of the short-range microstructural elements they contain, such as comonomers, regio- and stereosequences, branches, cross-links etc., as well as their locations along the polymer backbone. Consequently, the present situation regarding our ability to characterize the complete chemical structures of synthetic polymers would be analogous to that of proteins if it were only possible to determine their amino acid compositions or possibly the amounts of consecutive pairs or even triplets of constituent amino acids.

Reading the DNA genome used to synthesize proteins generally enables us to know their complete macrostructures, i.e., their complete amino acid sequences or primary structures. Just as protein secondary, tertiary, and even quaternary structures, and of course their resultant biological functions, are determined by their amino acid sequences (see Chapter 7), the behaviors of synthetic polymers can logically be presumed to also principally result from their complete structural architectures. Though important, knowledge of the types and quantities of short-range microstructures that polymers contain, which constitutes our present level of structural knowledge, is insufficient for the development of truly relevant structure-property relations.

A related issue also needs to be addressed. The degree of macrostructural heterogeneity among the chains in polymer samples is also expected to strongly influence the behaviors of materials made from them. Our attempts to develop and demonstrate an experimental approach that can be used to begin to characterize the complete macrostructures of synthetic polymers and to illustrate the relevance of this knowledge to understanding their properties and behavior have been recently reviewed (Gurarslan and Tonelli 2017).

Here, we will briefly summarize the Kerr effect characterization of polymers and provide a few example applications.

INTRODUCTION

A potential experimental probe to monitor or characterize polymer macrostructures must have an observable output that depends on the entire polymer chain. Overall dimensions ($<r^2>$ or $<s^2>$) are such a property. However, the structural sensitivity of chain dimensions to the locations of constituent short-range microstructures is not sufficient to characterize polymer macrostructures. Recently, we summarized our investigations of polymer microstructures and conformations through use of the contributions made by polymer solutes to the birefringence observed in their dilute solutions (Δn) when subjected to strong electric fields (E) (see Figure A.1), i.e., their Kerr effects (Kerr 1875, 1879, 1880, 1882, 1894a, 1894b; Tonelli 2009). We concluded there that the microstructures and the resultant conformations of polymers sensitively affect their observed Kerr effects. For example, we noted that distinct Kerr effects are produced for polymers possessing the same microstructures, but that are located differently along their chains, and might be able to distinguish different overall polymer chain architectures or macrostructures. Gurarslan and Tonelli, 2017.

Molecular Kerr effects, including those of conformationally flexible molecules like polymers, depend on the magnitudes and orientations of their overall net dipole moments and anisotropic polarizability tensors (see Figure A.1).

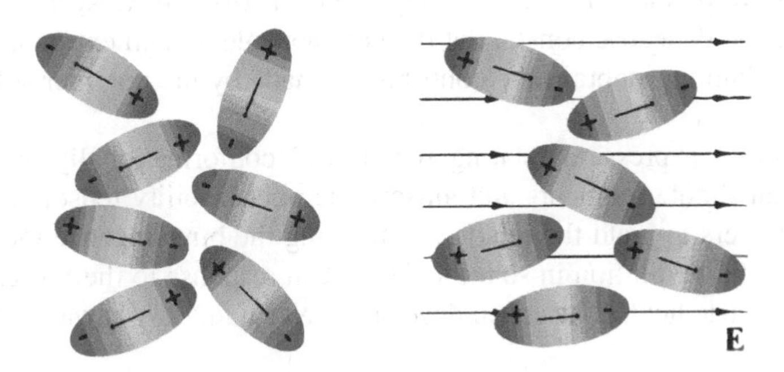

FIGURE A.1 Schematic diagram illustrating the dipolar alignment of molecules in an electric field and the resultant birefringence that is produced by the partial alignment of their net overall polarizability tensors. (From Gurarslan, R. and Tonelli, A. E., *Prog. Polym. Sci.*, 65, 42–52, 2017.)

The Kerr Effect

The induction of birefringence in an isotropic sample by application of an electric field.

John Kerr (1875)

The Kerr effect may be sensitive enough to characterize the macrostructures of polymers, because the molar Kerr constants ($_mKs$) of small molecules, ~the sizes of monomers, are seen to vary over nearly five orders of magnitude and may be either positive or negative (Bulgarevich and Burdastykha 2008).

The RIS conformational description of polymer chains, in combination with the assumption of the additivity of bond dipole moments and polarizability tensors, and utilization of matrix multiplication methods developed to rigorously sum and average them over all realistic conformations available to polymers, allow the estimation of the molar Kerr constants of polymer chains (Flory 1969). This is critical, because only from the successful comparison of $_mKs$ calculated for polymers, with assumed macrostructures that are in agreement with observed Kerr effects, can we conclude that the assumed and actual macrostructures are likely the same.

The experimental determination of the $_mKs$ of polymer solutes have been summarized (Briegleb 1931; Otterbein 1933, 1934; Sachsse 1935; Friedrich 1937; Riande and Saiz 1992). The $_mKs$ of a polymer solute obtained from the electrical birefringence observed at infinite dilution in solutions with isotropic solvents may be experimentally derived from the following relation (LeFevre and LeFevre 1955, 1960):

$$_m\mathbf{K} = (\mathbf{6N_A\lambda nB}) / \left[\rho(\mathbf{n} + \mathbf{2})^2 (\varepsilon + \mathbf{2})^2\right],$$

where **n**, **B**, **ρ**, and **ε** are, respectively, the refractive index, Kerr constant, density, and dielectric constant of the polymer solution, all extrapolated to infinite dilution to obtain the contribution made by the polymer solute to each.

A formal expression relating the overall conformationally averaged molecular dipole moments and anisotropic polarizability tensors of flexible polymers to yield the relation connecting the birefringence they produce at infinite dilution in solution, i.e., $_mK$ in response to their alignment by the electric field E, has been developed (Nagai and Ishikawa 1965).

$$_m\mathbf{K} = \left(\mathbf{2\pi N_A / 135}\right)\left[(<\mu^T\alpha\mu>)/\mathbf{k^2T^2} + (<\alpha^R \acute{\alpha}^C>)/\mathbf{kT}\right].$$

$\boldsymbol{\mu}^T$ and $\boldsymbol{\mu}$ are the overall polymer dipole moments expressed as a row or column vector, respectively. $\boldsymbol{\alpha}$ and $\boldsymbol{\acute{\alpha}}$ are the overall anisotropic optical and static polarizability tensors of the polymer, respectively, with both contributing to $\mathbf{\Delta\alpha}$, the overall difference in polarizabilities of the entire polymer chain in directions parallel and perpendicular to the applied field **E**. Their superscripts **R** and **C** designate row and column forms of their traceless tensors, and $< >$ indicates the appropriate average of both the dipolar and polarizability contributions over all conformations available to the polymer chain. $\mathbf{N_A}$ and **k** are the Avogadro and Boltzmann constants, respectively, and **T** is the temperature.

To perform the correct averaging of both the dipolar $\langle \mu^T \alpha \mu \rangle$ and polarizability $\langle \alpha^R \acute{\alpha}^C \rangle$ terms of the Nagai-Ishikawa expression for $_mK$ over all of a polymer chain's realistic RIS conformations, Flory and Jernigan developed the appropriate matrix multiplication techniques (Flory and Jernigan 1968; Flory 1969, 1974). Their matrix method assumes constituent bond dipole moments and polarizability tensors that are additive and independent of the bond conformation, making possible a comparison of experimentally observed molar Kerr constants for polymers with those realistically estimated (see Appendix 4.2 where a Fortran program for calculating the Kerr constants of polymers is presented). Their development was critical to the determination of both the underlying micro- and macrostructures of polymers with the Kerr effect.

The constituent short-range microstructures of polymers may be arranged along their chains in an astoundingly large number of ways. Thus, calculation of an individual $_mK$ for each potential macrostructure is precluded (Saiz et al. 1977; Tonelli 1977). For example, a 50:50 A/B copolymer sample containing chains each with 100 total comonomers can potentially have 100!/50!50! or ~10^{29} A/B copolymer chains with different comonomer sequences. We suggest using the currently best method, high resolution solution ^{13}C-NMR, for determining the types and amounts of constituent short-range microstructures present in polymers, and then use the Kerr effect to locate these constituent microstructures along the backbone of the chain. When an estimated $_mK$ based on our assumed locations of the NMR-derived microstructures agrees with the observed value, then it is likely that the assumed and actual polymer chain macrostructures are closely similar.

Now we summarize and provide several examples of using the Kerr effect to begin to successfully analyze/characterize polymer macrostructures and indicate the profound effects they can have on the properties and behaviors of materials made from them.

STYRENE/p-BRSTYRENE COPOLYMERS

Several important attributes of styrene/p-BrStyrene (S/pBr) copolymers influenced our decision to examine them in detail with the Kerr effect (Khanarian et al. 1982; Semler et al. 2007; Hardrict et al. 2013;

Gurarslan et al. 2015a, 2015b; Gurarslan and Tonelli 2016). First, the net dipole moments and polarizability tensors of S and pBrS repeat units are quite different. Second, the randomly coiling conformations for all S/pBrS copolymers with the same stereosequences are essentially identical (Yoon et al. 1975). That is why portions of their ^{13}C-NMR spectra potentially sensitive to their microstructures, i.e., the resonance frequencies of their backbone CH and CH_2 and side-chain C_1 carbons, are independent of both their comonomer compositions and sequences. This renders their attempted determination by ^{13}C-NMR relatively useless. Third, the calculation of $_m$Ks is greatly simplified by copolymer conformations that are dependent solely on stereosequences. This latter reason is especially important, because $_m$Ks measured for polymers are interpretable, in terms of the types, quantities, and the locations of their microstructures, i.e., their macrostructures, only by comparing them with the $_m$Ks calculated for the corresponding micro- and macrostructural features.

In Figure A.2, we display molar Kerr constants calculated for S/pBrS copolymers with random comonomer sequences and various random stereosequences and compare them to those of S/pBrS copolymers produced by the random bromination of atactic polystyrene (a-PS). The impetus for this Kerr effect investigation was provided by the fact that at that time the tacticity of a-PS was in question (Khanarian et al. 1982). Note that only the $_m$Ks calculated assuming that the starting a-PS and consequently the resultant S/pBrS copolymers have a tacticity characterized by a random distribution of 55% racemic (*r*) and 45% meso (*m*) diads agree closely with the observed $_m$Ks. Subsequent ^{13}C-NMR observations of various stereoisomeric pentad model compounds of a-PS unambiguously confirmed this conclusion (Sato et al. 1982, 1983).

In Table A.1, molar Kerr constants calculated and observed for atactic S/pBrS copolymers were also made by the bromination of a-PS, but in solvents of different quality for a-PS. (Semeler et al. 2007). The θ–temperatures of a-PS are 6.6°C, 32.8°C, and 58.6°C, respectively, in 1-chlorodecane (CD), 1-chloroundecane (CUD), and 1-chlorododecane (CDD). At 33°C, bromination of a-PS in CD, CUD, and CDD would be expected to produce increasingly blocky S/pBrS copolymers, as the a-PS coils contract. As demonstrated in Table A.1, this expectation is confirmed by comparison of calculated and observed molar Kerr constants.

A final and extreme example of the sensitivity of the Kerr effect to the comonomer sequences of S/p-BrS copolymers is provided by the triblock copolymers shown in Figure A.3. Both pBrS90-b-S120-b-pBrS90 (I) and S60-b-pBrS180-b-S60 (II) were synthesized by Reversible addition-fragmentation chain transfer (RAFT) polymerization (Hardrict et al. 2013). Because only very small quantities of these triblock

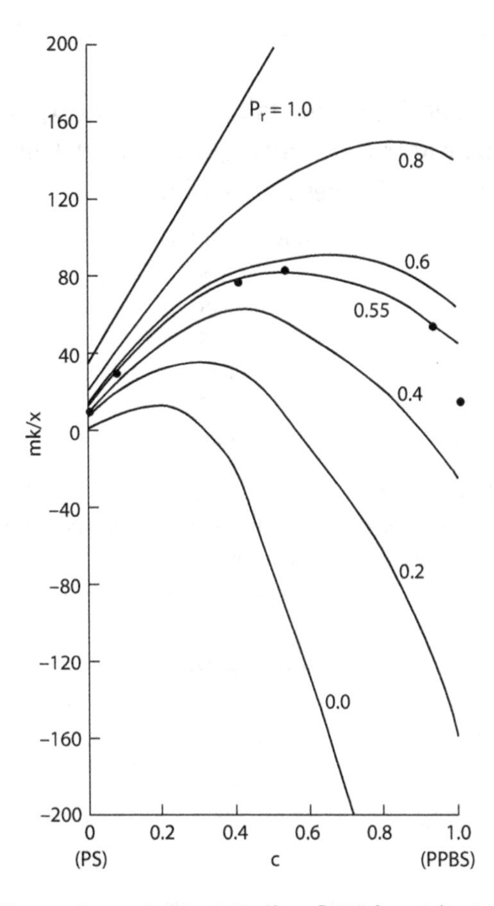

FIGURE A.2 Comparison of $_m$Ks (×10^{-12} cm^7 SC^{-2} mol^{-1}) observed in dioxane (•) and calculated (curves) for atactic S/pBrS copolymers, with random comonomer sequences. X, P_r, and c are, respectively, the number of repeat units (200), fraction of racemic diads, and fractional contents of pBrS units. (Reprinted with permission from Khanarian, G. et al., *Macromolecules*, 15, 866, 1982. Copyright 1982 American Chemical Society.)

copolymers were produced, measurement of the electrical birefringence for a complete series of their dilute dioxane solutions, necessary to determine their molar Kerr constants, was not possible.

Instead, measurements were restricted to one or two solutions at low concentrations (<1.0 wt%) for each triblock. Nevertheless, the $_m$Ks for both triblocks appear to have similar magnitudes, but with $_m$K(I) negative and K(II) positive, based on these preliminary partial electrical birefringence observations. The birefringence of 0.6 wt% solutions were observed to decrease (I) and increase (II), respectively, the birefringence of their dioxane solutions by 17% and 26%. Thus, triblock (I) likely has a negative $_m$K, whereas the $_m$K for triblock (II) is positive.

TABLE A.1

Calculated (RIS) and Experimentally Measured (Exp) and Normalized Molar Kerr Constants ($_mK$) of S/pBrS Copolymers with Different Comonomer Sequence Distributions

Monomer Sequence	Determination of $_mK$	$_mK/_mK$ Random
$(PS_2$-b-$PBrS_3)_{60}$	RIS	0.773
$(PS_4$-b-$PBrS_6)_{30}$	RIS	0.621
$(PS_{10}$-b-$PBrS_{15})_{12}$	RIS	0.363
$(PS_{20}$-b-$PBrS_{30})_6$	RIS	0.216
$PBr_{0.63S}$-CUD_{33} ($T = \theta$)	Exp	0.533[a]
$PBr_{0.58}S$-CDD33 ($T > \theta$)	Exp	0.304[a]

Source: Semler, J. J. et al., *Adv. Mater.*, 19, 2877, 2007.

[a] Normalized with the molar Kerr constant measured for $PBr_{0.59}S$-CD33 ($T > \theta$), having an expected more random sequence of comonomers.

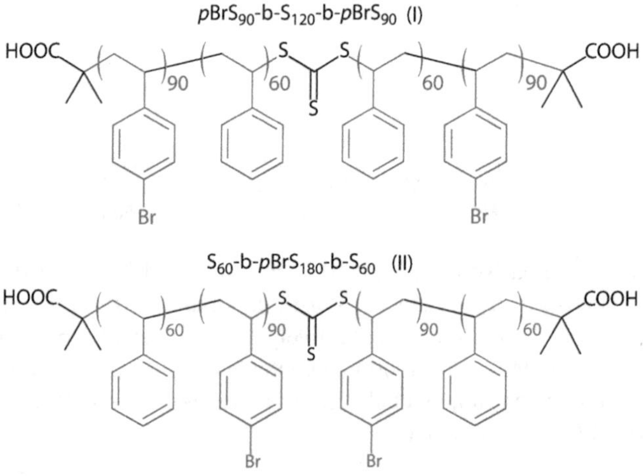

FIGURE A.3 Structures of $pBrS_{90}$-b-S_{120}-b-$pBrS_{90}$ (I) and S_{60}-b-$pBrS_{180}$-b-S_{60} (II) triblock copolymers synthesized by RAFT copolymerization. (Reprinted with permission from Hardrict, S. N. et al., *J. Polym. Sci., Polym. Phys.*, 51, 735, 2013.)

Even though both triblock copolymers have the same overall comonomer composition and number of identical block junctions, they evidence Kerr effects with opposite signs, which are consistent with their calculated values. Short-range local probes, such as ^{13}C-NMR spectroscopy cannot distinguish between these S/pBrS triblock copolymers, while the Kerr effect can. Because they differ only in the locations of their S and pBrS blocks, it is clear the Kerr effect cannot only identify polymer microstructures, but also can locate them along the macromolecular chain.

Random atactic 20:80, 50:50, and 80:20 S/pBrS copolymers were also obtained *via* un-controlled free-radical copolymerization (Gurarslan et al. 2015b). The 50:50 copolymer was expected to have a larger molar Kerr constant than those of both the 20:80 and 80:20 copolymers, as is observed and calculated for a-S/pBrS copolymers (see Figure A.2). Though with the same overall comonomer concentration, a heterogeneous 1:1 mixture of the 20:80 and 80:20 copolymers was also expected and observed to have a Kerr constant less than the 50:50 copolymer, demonstrating the capability of Kerr effect observations to distinguish polymer samples with a heterogeneous distribution of macrostructures within and among their constituent chains. The Kerr effect appears to be able to distinguish between samples whose polymer chains all have the same macrostructures from those with a heterogeneous distribution, but with the same overall sample average of local microstructures. Of course, this distinction cannot be made by spectroscopies whose sensitivities are limited to short-range microstructures like NMR.

To emphasize the limitations of ^{13}C-NMR and the potential of the Kerr effect to unravel the complete architectures or macrostructures of synthetic polymers, we offer the cartoon in Figure A.4. Need we say more?

FIGURE A.4 Cautionary note: Observation tools can bias structural perspectives.

Properties

The glass transition temperatures (T_gs) of atactic S/pBrS copolymers are dependent on comonomer composition, but are observed to be independent of comonomer sequence, likely because they essentially have the same conformational characteristics (Yoon et al. 1975; Semler et al. 2007; Tonelli et al. 2010). On the other hand, we would expect their T_gs to depend on stereosequence, because their conformational characteristics do (Gurarslan and Tonelli 2016). As is apparent in Figure A.5, T_gs of atactic S/pBrS copolymers obtained by bromination of a-PS [Molecular Weight (MW)] = 30,000 in solvents of different quality for a-PS do depend only on commoner composition.

Figure A.6 summarizes the dewetting behavior of these same random and blocky atactic S/pBrS copolymers. In good/poor solvents, the a-PS coils are expanded/contracted leading to random/blocky bromination. It is clear from these results that the high temperature dewetting of atactic S/pBrS thin films depends on the distributions or sequences of comonomers, as well as their compositions (though not shown).

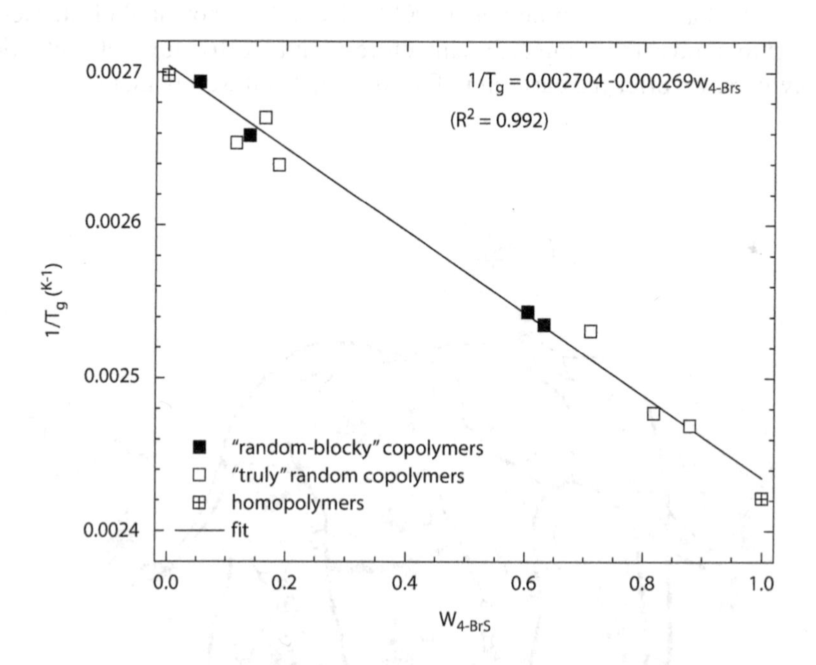

FIGURE A.5 The glass-transition temperature for atactic S/pBrS copolymers of various comonomer compositions (weight fractions) and sequences. (Reprinted with permission from Tonelli, A. E. et al., *Macromolecules*, 43, 6912, 2010. Copyright 2010 American Chemical Society.)

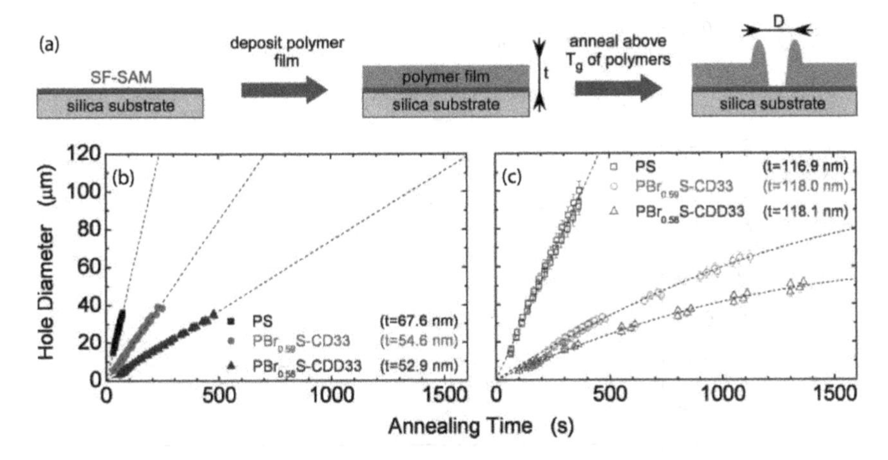

FIGURE A.6 (a) Samples for dewetting studies were formed by covering the silicon wafers with a self-assembled monolayer of 1H,1H,2H,2H-perfluorodecyltrichlorosilane (SF-SAM), casting thin polymer films (thickness t), and heating to temperatures *T*, such that $T - T_g = 35°C$. Upon annealing, holes with diameter *D* develop in the films and grow with increasing time. (b, c) Hole diameters for PS (squares), $PBr_{0.59}$S-CD33 (circles), and PBr_{58}-S-CDD33 (triangles) measured on films with $t \approx 60$ and 117. (With permission Reproduced from Semler, J. J. et al., *Adv. Mater.*, 19, 2877, 2007.)

STYRENE/BUTADIENE MULTIBLOCK-COPOLYMERS

Kerr Effects

The random and regularly alternating styrene/butadiene (S/B) multiblock copolymers shown below (see Figure A.7) were synthesized by Lee and Bates. Both contain ~0.8 volume fraction of S units, but showed distinct Kerr constants, while their ^{1}H-NMR spectra were indistinguishable (Lee and Bates 2013).

Properties

As evidenced by Differential Scanning Calorimetry (DSC), both the random and regularly alternating multiblock samples showed phase segregation, with T_gs for regularly alternating and random multiblock copolymers, respectively, of 81°C and 90°C for the S-rich hard phases and −90°C and −92°C for the B-rich soft phases (Lee and Bates 2013). This comparison was made between random and regularly alternating multiblock copolymers containing a volume fraction of S of ~0.8,

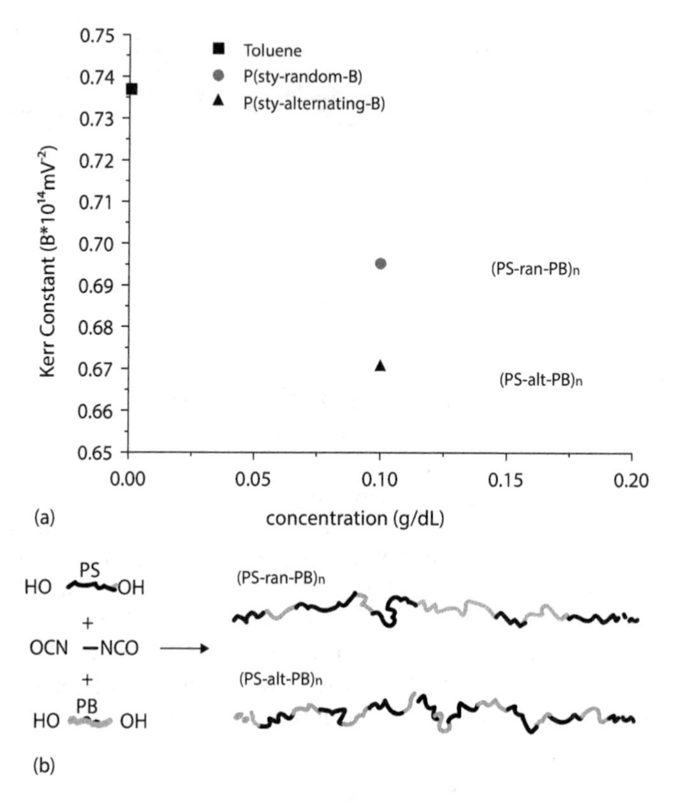

FIGURE A.7 (a, b) Structures and Kerr constants of 0.1 g/dL toluene solutions of 80/20 styrene/butadiene multiblock copolymers with random and regularly alternating blocks. (Reprinted with permission from Lee, I. and Bates, F. S., *Macromolecules*, 46, 4529, 2013. Copyright 2013 American Chemical Society; From Gurarslan, R. et al., *J. Polym. Sci., Polym. Phys.*, 53, 155, 2015.)

using the same S and B blocks. In both cases, each phase contained ~96% of one or the other blocks.

A single broad diffraction peak was revealed by Small Angle X-ray Scattering (SAXS) observations and, as expected, showed larger domain sizes for the random multiblock samples with larger average block lengths (Lee and Bates 2013). Transmission electron microscopy (TEM) micrographs showed an irregular, disordered, bicontinuous-like morphology, rather than the ordered-phase-separated morphologies that were evidenced in the TEM micrographs of di and triblock copolymers of S and B (Lee and Bates 2013).

In uniaxial tests of rectangular bars of S/B multiblocks, the following behaviors were seen: random multiblocks had higher elastic moduli, lower elongations at break, and similar yielding and breaking stresses compared to the regular alternating multiblocks (Lee and Bates 2013). In Figure A.8,

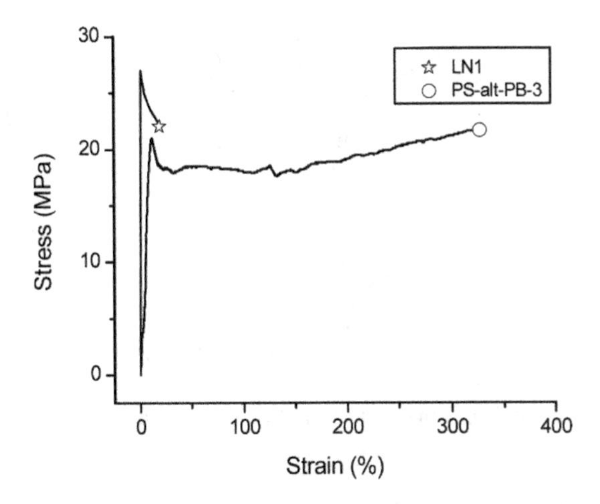

FIGURE A.8 Comparison of the tensile behavior of similar composition and molecular weight SBS triblock (LN1) and PS-alt-PB-3 S-B multiblock copolymers. (Reprinted with permission from Lee, I. and Bates, F. S., *Macromolecules*, 46, 4529, 2013. Copyright 2013 American Chemical Society.)

more remarkable are the very different mechanical behaviors shown by a SBS triblock and an S/B multiblock with the same composition and molecular weight. While the SBS triblock is significantly stiffer, with a higher elastic modulus and evidences no necking, the S/B multiblock is much tougher, with more than an order of magnitude larger strain at break. Morphological distinctions between these block copolymer samples are clearly the source for these differences. The triblock, with discontinuous soft B cylinders dispersed in a continuous hard S matrix, fractures rather than necks upon stretching. On the other hand, the high S content glassy portions in the S/B multiblock easily fragment and become embedded in a soft rubbery high B content continuous domain, causing a plastic-to-rubber transition, which leads to necking and produces a much tougher material.

Before concluding this Appendix describing the use of the Kerr effect to locate the positions of polymer microstructures along and among their chains, we want to illustrate that the Kerr effect is also very sensitive to local short-range polymer chain conformations.

Two RIS conformational models were simultaneously developed for poly(vinyl bromide) (PVB), one by Saiz et al. and the other by one of us (Saiz et al. 1982; Tonelli 1982). Though each was successful in reproducing the dimensions and dipole moments measured for atactic-PVB at room temperature, they were different in detail. Each found the *tg* and *gt* conformations to be preferred for *m*-PVB diads, but with different populations. For racemic *r*-PVB diads, both RIS models found the *tt* and gg

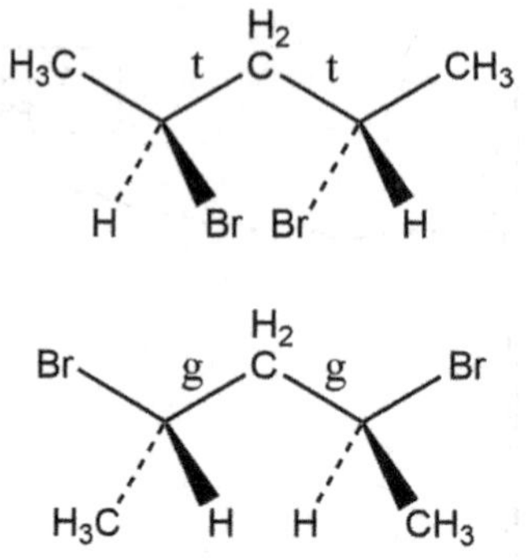

FIGURE A.9 Schematic representation of r-DBP in the tt and gg conformations.

conformers to be preferred. However, RIS-T predicted the *gg* conformer to be preferred over the *tt* conformer, while the RIS-S RIS model indicated the opposite preference (see Figure A.9).

In an effort to further test the two PVB RIS models, the stereoisomeric PVB oligomers *m*- and *r*-dibromopentanes (DBPs) and *mm*-, rr-, and *mr*-tribromoheptantes (TBHs) were synthesized, and their dipole moments and molar Kerr constants were measured and calculated with both RIS models (Tonelli et al. 1985). These results are presented in Table A.2 and show that both RIS models reproduce the measured dipole moments.

TABLE A.2
Dipole Moments (D^2) and Molar Kerr Constants ($\times 10^{-12}$ cm^{-7} SC^{-2} mol^{-1}) for the Stereoisomers of DBP and TBH

Isomer		$\langle \mu^2 \rangle$			$_mK$	
m	5.29[a]	5.46[b]	5.27[c]	8.46[a]	9.22[b]	10.5[c]
r	5.00	5.64	5.52	13.6	14.9	9.3
M[d]	5.06	5.56	5.41	10.8	12.5	9.8
mm	—[a]	4.68[b]	4.30[c]	—[a]	−82[b]	−77[c]
mr	—	6.08	8.04	—	−28	58
rr	—	12.7	12.1	—	267	245
M[d]	8.58	8.41	9.05	77.6	76.5	111

Source: Tonelli, A. E., *NMR Spectroscopy and Polymer Microstructure: The Conformational Connection*, VCH, New York, 1989.

[a] Measured in CCl_4 at 25°C.

[b] Calculated using the RIS-T conformational model.

[c] Calculated using the RIS-S conformational model.

[d] 58:42 *r:m* mixture of DBP and 38:48:14 *rr:mr:mm* mixture of TBH isomers.

In contrast, only the RIS-T conformational model leads to calculated $_{m}$Ks in agreement with the measured values for both DBP and TBH stereoisomeric oligomers.

The reason for this behavior can be understood by considering the preferred *tt* and *gg* conformations for r-DBP shown in Figure A.9. The net dipole moment and the plane of maximum polarizability, due to the presence there of both highly anisotropic C-Br bonds, are coincident with the plane of the backbone in the *gg* conformer. The C-Br bonds, as well as their net dipole moment and planes of maximum polarizability, are nearly perpendicular in the *tt* backbone conformer, so the overall anisotropy of the polarizability tensor is less than that in the *gg* conformation. The *gg* conformer has a calculated $_{m}K = 20$, while $_{m}K = 2$ for the *tt* conformer.

In the RIS-S model, the *tt* conformer is preferred over the *gg* conformer, while the RIS-T model predicts the opposite conformational preference. This explains why $_{m}K = 9$ and 15 when calculated for r-DBP with the RIS-S and RIS-T models, respectively, compared to the measured $_{m}K = 14$. The high sensitivity of $_{m}K$ to both the conformational and configurational characteristics of flexible polymers like PVB is illustrated by the discrimination and selection among these two RIS models that was afforded by comparison of their predicted $_{m}$Ks to the observed values.

SUMMARY AND CONCLUSIONS

We believe that coupling Kerr effect observations of dilute polymer solutions with their ^{13}C-NMR spectra can usefully be applied to begin to determine their overall macromolecular architectures, i.e., their macrostructures. The types and quantities of local microstructures present in polymers can be obtained with high resolution ^{13}C-NMR observations of their solutions. The identified short-range microstructures can then be moved along and located on the polymer backbone until their resultant calculated molar Kerr constants reproduce those observed. When the assumed locations of the NMR-derived microstructures produce an estimated $_{m}K$ in agreement with the observed value, then it is likely that the assumed and actual polymer chain macrostructures are closely similar.

By way of several examples, we have demonstrated the ability of Kerr effect observations to distinguish between polymers composed of the same repeat units and with the same quantities, but that are molecularly incorporated to yield different overall architectures or macrostructures. To explicitly identify such distinct macrostructures, we must be able to estimate the molar Kerr constants expected for each, as was done for the

S/pBrS copolymers discussed here. This requires knowledge of the conformational characteristics (RIS models) of each distinct macrostructure, and entails, for example, development of nearest neighbor-dependent statistical weight matrices appropriate for *r* and *m* diads and AA, BB, BA or possibly even AAA, ABA, AAB, BBB, BAB, and BBA comonomer sequences. If RIS models are available for the poly A and poly B homopolymers, then development of RIS models for their copolymers is not difficult. In addition, estimates of constituent bond dipole moments and polarizability tensors are also necessary (Khanarian et al. 1982). However, in many instances, these are available for small molecules containing similar chemical bonds.

Without minimizing the effort inherent in reliably estimating the $_{m}$Ks of polymers to help determine their macrostructures, it must be realized that the behaviors of their resultant materials cannot be meaningfully understood without this complete structural knowledge. As in many other "real life" situations, more often than not, the expected value of results produced justify the labor required to obtain them.

The few examples discussed here serve to once again clearly illustrate that the behaviors of polymer materials are directly determined by the detailed molecular architectures/macrostructures of their constituent polymer chains. Coupling ^{13}C-NMR spectra and Kerr effect observations and analyzing them with their conformational characteristic/RIS models may be useful in the characterization of overall polymer architectures, i.e., their macrostructures, and eventually lead to the development of more relevant structure-property relations.

REFERENCES

Briegleb, G. Z. (1931), *Phys. Chem., B*, 14, 97–121.

Bulgarevich, S. B., Burdastykha, T. V. (2008), *Russian J. Gen. Chem.*, 78, 1307.

Flory, P. J., Jernigan, R. L. (1968), *J. Chem. Phys.*, 48, 3823.

Flory, P. J. (1969), *Statistical Mechanics of Chain Molecules*, Wiley-Interscience, New York, Chapter 9.

Flory, P. J. (1974), *Macromolecules*, 7, 381.

Friedrich, H. (1937), *Phys. Z.*, 38, 318.

Gurarslan, R., Hardrict, S. N., Roy, D., Galvin, C., Hill, M. R., Gracz, H., Sumerlin, B. S., Genzer, J., Tonelli, A. E. (2015a), *J. Polym. Sci., Polym. Phys.*, 53, 155.

Gurarslan, R., Gurarslan, A., Tonelli, A. E. (2015b), *J. Polym. Sci., Part B: Polym. Phys*, 53, 409.

Gurarslan, R., Tonelli, A. E. (2016), *Polymer*, 89, 50.

Gurarslan, R., Tonelli, A. E. (2017), *Prog. Polym. Sci.*, 65, 42–52.

Hardrict, S. N., Gurarslan, R., Galvin, C. J., Gracz, H., Roy, D., Sumerlin, B. S., Genzer, J., Tonelli, A. E. (2013), *J. Polym. Sci., Polym. Phys.*, 51, 735.

Kerr, J. (1875), *Philos. Mag. Ser.*, 4, 50, 337, 446; (1879), *Philos. Mag. Ser.*, 5, 8 (85), 85, 229; (1880), *Philos.* Mag. Ser., 5, 9, 157; (1882), *Philos. Mag. Ser.*, 5, 13, 153, 248; (1894a), *Philos. Mag. Ser.*, 5, 37, 380; (1894b), *Philos. Mag. Ser.*, 5, 38, 1144.
Khanarian, G., Cais, R. E., Kometani, J. M., Tonelli, A. E. (1982), *Macromolecules*, 15, 866.
Lee, I., Bates, F. S., (2013), *Macromolecules*, 46, 4529.
LeFevre, C. G., LeFevre, R. J. W. (1955), *Rev. Pure Appl. Chem.*, 5, 261.
LeFevre, C. G., LeFevre, R. J. W. (1960), *Techniques of Organic Chemistry*, Weissberger, A., Ed.; Interscience, New York, Vol. I, Chapter XXXVI.
Nagai, K., Ishikawa, T. (1965), *J. Chem. Phys.*, 43, 4508.
Otterbein, G., (1933), *Phys. Z.*, 34, 645; (1934), 35, 249.
Riande, E., Saiz, E. (1992), *Dipole Moments and Birefringence of Polymers*, Prentice Hall, Englewood Cliffs, NJ, Chapter 7, 192–258.
Sachsse, G. (1935), *Phys. Z.*, 36, 357.
Saiz, E., Riande, E., Delgadp, M. P., Barrales-Rienda, J. M. (1982), *Macromolecules*, 15, 1152.
Saiz, E., Suter, U. W., Flory, P. J. (1977), *J. Chem. Soc., Faraday Trans.*, 2, 73, 1538.
Sato, H., Tanaka, Y., Hatada, K. (1982), *Makromol. Rapid Commun.*, 3, 181; (1983), *J. Polym. Sci., Part B: Polym. Phys Ed.*, 21, 1667.
Semler, J. J., Jhon, Y. K., Tonelli, A., Beevers, M., Krishnamoorti, R., Genzer, J. (2007), *Adv. Mater.*, 19, 2877.
Tonelli, A. E. (1977), *Macromolecules*, 10, 153.
Tonelli, A. E. (1982), *Macromolecules*, 15, 290.
Tonelli, A. E. (1989), *NMR Spectroscopy and Polymer Microstructure: The Conformational Connection*, VCH, New York.
Tonelli, A. E. (2009), *Macromolecules*, 42, 3830.
Tonelli, A. E., Jhon, Y. K., Genzer, J. (2010), *Macromolecules*, 43, 6912.
Tonelli, A. E., Khanarian, G., Cais, R. E. (1985), *Macromolecules*, 18, 2324.
Yoon, D. Y., Sundararajan, P. R., Flory, P. J. (1975), *Macromolecules*, 8, 77.

DISCUSSION QUESTIONS

1. Why do Kerr effect observations of dilute polymer solutions potentially provide an experimental means to more fully characterize polymer microstructures beyond those determined by ^{13}C-NMR?
2. How can the experimentally determined molar Kerr constant, $_{m}K$, of a polymer be connected to or identified with the macrostructure that produced it?
3. Why is it important to begin determining the complete architectures or macrostructures of synthetic polymers?
4. Why was it suggested that ^{13}C-NMR observation be used to determine the types and amounts of short-range microstructures present in synthetic polymers, while Kerr effect observations are subsequently used to locate their positions along the polymer chain?
5. How can the procedures for determining polymer macrostructures suggested in number 4 actually be carried out?

APPENDIX 4.2: PROGRAM (FORTRAN) USED TO CALCULATE MOLAR KERR CONSTANTS FOR POLYMERS

$$ {}_mK = \left(2\pi N_A/135\right)\left[\left(\left\langle \mu^T \alpha \mu \right\rangle\right)/k^2T^2 + \left(\left\langle \alpha^R \acute{\alpha}^C \right\rangle\right)/kT\right] $$

$$ \left\langle \mu^T \alpha \mu \right\rangle = 2Z^{-1}\mathrm{J}^*\mathrm{Q}_1^{n+1}\mathrm{J} $$

$$ \mathbf{Q}_i = \begin{bmatrix} U & (U\otimes 1^T)\|T\| & (U\otimes \hat{\alpha}^R)\|T\otimes T\| & \frac{1}{2}(U\otimes 1^T\otimes 1^T)\|T\otimes T\| & U\otimes\left[\hat{\alpha}^R(1\otimes E_3)\right]\|T\| & \frac{1}{2}U\left[\hat{\alpha}^R(1\otimes 1)\right] \\ 0 & (U\otimes E_3)\|T\| & 0 & (U\otimes E_3\otimes 1^T)\|T\otimes T\| & (U\otimes\hat{\alpha})\|T\| & U\otimes\left[(E_3\otimes 1^T)\,\hat{\alpha}^C\right] \\ 0 & 0 & (U\otimes E_9)\|T\otimes T\| & 0 & (U\otimes 1\otimes E_3)\|T\| & \frac{1}{2}(U\otimes 1\otimes 1) \\ 0 & 0 & 0 & (U\otimes E_9)\|T\otimes T\| & 0 & U\otimes\hat{\alpha}^C \\ 0 & 0 & 0 & 0 & (U\otimes E_3)\|T\| & U\otimes 1 \\ 0 & 0 & 0 & 0 & 0 & U \end{bmatrix}_i $$

$\mathbf{Q}_i$ is $26\nu \times 26\nu$, where ν is the number of conformations permitted about each backbone bond (usually 3), J^* and J are $1 \times 26\nu$ and $26\nu \times 1$ vectors, where, respectively, the first and last three elements are 1s, and the bond vector **l** should be replaced by the dipole moment vector **m**.

$$ \left\langle \alpha^R \acute{\alpha}^C \right\rangle = 2Z^{-1}\mathrm{J}^*\mathrm{A}_1^{n+1}\mathrm{J} $$

$$ \mathbf{A}_i = \begin{bmatrix} U & \left(U\otimes\hat{\alpha}^R\right)\|T\otimes T\| & \left(\hat{\alpha}^R\hat{\alpha}'^C\right)U \\ 0 & \left(U\otimes E_3\right)\|T\otimes T\| & U\otimes\hat{\alpha}'^C \\ 0 & 0 & U \end{bmatrix}_i $$

A_i is $11\nu \times 11\nu$, J^* and J are $1 \times 11\nu$ and $11\nu \times 1$ vectors, where, respectively, the first and last three elements are 1s (Flory 1969).

This program, written in FORTRAN, calculates the $_m$Ks for 300 repeat unit chains of the triblock BrS_{90}–S_{120}–BrS_{90} each with a tacticity of p_r = 0.51, 0.52, 0.53, or 0.54. 200 sample chains were generated for each P_r to obtain the <$_m$Ks>s.

```
  real t1(3,3),t2(3,3),ph1,ph2,s1,s2,c1,c2,z,
 7up(2,2),uppm(2,2),uppr(2,2),eta, w,wp, wpp,
 7u1(2,2),u2(2,2),temp, pr, rn, dorl, pbrs,
 7mubrsl(3,1),mubrsd(3,1),mutbrsl(1,3),mutbrsd(1,3),
 7musl(3,1),musd(3,1),mutsl(1,3),mutsd(1,3),
 7acapsl(3,3),arsl(1,9),acsl(9,1),acapsd(3,3),arsd(1,9),
 7acsd(9,1),acapbrsl(3,3),arbrsl(1,9),acbrsl(9,1),
 7acapbrsd(3,3),arbrsd(1,9),acbrsd(9,1),e3(3,3),e9(9,9),
```

```
7tbarsl(6,6),tbarsd(6,6),tbarsll(6,6),tbarsdl(6,6),
7tbarsld(6,6),tbarsdd(6,6),txtbarsl(18,18),
7txtbarsd(18,18),txtbarsdd(18,18),txtbarsll(18,18),
7txtbarsdl(18,18),txtbarsld(18,18),txt(9,9),asl(22,22),
7asd(22,22),asdd(22,22),asll(22,22),asdl(22,22),
asld(22,22),
7qsl(52,52),qsd(52,52),qsll(52,52),qsdd(52,52),
qsdl(52,52),
7qsld(52,52),abrsl(22,22),abrsd(22,22),qbrsl(52,52),
7qbrsd(52,52),uxe9(18,18),uett(18,18),uxa(2,18),
uatt(2,18),
7arac, aau(2,2),uxac(18,2),uxe3(6,6),umuxe(18,6)
real uet(6,6),muxmu(1,9),uxmuxmu(2,18),muxe(9,3),
amue(1,3),
7uamue(2,6),uamuet(2,6),umumu(2,18),uuutt(2,18),
7mumu(9,1),armumu, uamumu(2,2),exmut(3,9),
uexmut(6,18),
7uemutt(6,18),uacap(6,6),uat(6,6),emuac(3,1),
uemuac(6,2),
7umuxet(18,6),mumumu(18,2),uxmu(6,2),j(1,22),
js(22,1),jj(1,52),7jjs(52,1),ja(1,22),jaa(1,22),
jjq(1, 52),jjqq(1,52),jaajs,
7jjuaujjs, ke, keuau, keaa, numbrs, brs, ketot,
kaatot, kuautot
integer ij, ji, ijk, jki, kij, dum
```

The above portion declares all the variables used as to type (real or integer and scalar, vector, or tensor) and also specifies the dimensions of the vector and tensor variables.

```
100 format(3e16.5)
111 format(e16.5)
201 format(f10.3)
```

These three statements specify the formats used for printing out results

```
temp=298.16
print 201, temp
eta=0.8*(2.71828**(200.0/temp))
w=1.3/(2.71828**(1000.0/temp))
wp=w
wpp=1.8/(2.71828**(1100.0/temp))
up(1,1)=1.0
up(1,2)=1.0
up(2,1)=1.0
```

```
up(2,2)=0.0
uppm(1,1)=wpp
uppm(1,2)=1.0/eta
uppm(2,1)=1.0/eta
uppm(2,2)=w/(eta*eta)
uppr(1,1)=1.0/1.1
uppr(1,2)=(wp/eta)/1.1
uppr(2,1)=(wp/eta)/1.1
uppr(2,2)=1.0/((eta*eta)*1.1)
```

The above are the *μ(i,j)* statistical weight elements in the statistical weight matrices *U(ij)* necessary for averaging over all polymer conformations.

```
t1(1,1)=0.37461
t1(1,2)=0.92718
t1(1,3)=0.0
t2(1,1)=0.40674
t2(1,2)=0.91355
t2(1,3)=0.0
```

t(a, b)s are the first row elements in the Euler matrix that transforms a vector or tensor from backbone bond i to the previous backbone bond i-1

```
mubrsl(1,1)=-0.63
mubrsl(2,1)=0.89
mubrsl(3,1)=1.54
mutbrsl(1,1)=-0.63
mutbrsl(1,2)=0.89
mutbrsl(1,3)=1.54
mubrsd(1,1)=-0.63
mubrsd(2,1)=0.89
mubrsd(3,1)=-1.54
mutbrsd(1,1)=-0.63
mutbrsd(1,2)=0.89
mutbrsd(1,3)=-1.54
musl(1,1)=-0.107
musl(2,1)=0.151
musl(3,1)=0.261
mutsl(1,1)=-0.107
mutsl(1,2)=0.151
mutsl(1,3)=0.261
musd(1,1)=-0.107
musd(2,1)=0.151
musd(3,1)=0.261
mutsd(1,1)=-0.107
mutsd(1,2)=0.151
mutsd(1,3)=0.261
```

The above are the dipole moments of the p-BrS phenyls attached on either side of the backbone (d or l) expressed either as a column (mu) or a row (mut).

```
acapsl(1,1)=-1.4045
acapsl(1,2)=-1.7002
acapsl(1,3)=-0.7838
acapsl(2,1)=-1.7002
acapsl(2,2)=-0.2022
acapsl(2,3)=1.1085
acapsl(3,1)=-0.7838
acapsl(3,2)=1.1085
acapsl(3,3)=1.6066
arsl(1,1)=-1.4045
arsl(1,2)=-1.7002
arsl(1,3)=-0.7838
arsl(1,4)=-1.7002
arsl(1,5)=-0.2022
arsl(1,6)=1.1085
arsl(1,7)=-0.7838
arsl(1,8)=1.1085
arsl(1,9)=1.6066
acsl(1,1)=-1.4045
acsl(2,1)=-1.7002
acsl(3,1)=-0.7838
acsl(4,1)=-1.7002
acsl(5,1)=-0.2022
acsl(6,1)=1.1085
acsl(7,1)=-0.7838
acsl(8,1)=1.1085
acsl(9,1)=1.6066
acapsd(1,1)=-1.4045
acapsd(1,2)=-1.7002
acapsd(1,3)=0.7838
acapsd(2,1)=-1.7002
acapsd(2,2)=-0.2022
acapsd(2,3)=-1.1085
acapsd(3,1)=0.7838
acapsd(3,2)=-1.1085
acapsd(3,3)=1.6066
arsd(1,1)=-1.4045
arsd(1,2)=-1.7002
arsd(1,3)=0.7838
arsd(1,4)=-1.7002
arsd(1,5)=-0.2022
arsd(1,6)=-1.1085
arsd(1,7)=0.7838
```

```
arsd(1,8)=-1.1085
arsd(1,9)=1.6066
acsd(1,1)=-1.4045
acsd(2,1)=-1.7002
acsd(3,1)=0.7838
acsd(4,1)=-1.7002
acsd(5,1)=-0.2022
acsd(6,1)=-1.1085
acsd(7,1)=0.7838
acsd(8,1)=-1.1085
acsd(9,1)=1.6066
acapbrsl(1,1)=-2.2823
acapbrsl(1,2)=-2.5331
acapbrsl(1,3)=-1.4914
acapbrsl(2,1)=-2.5331
acapbrsl(2,2)=-0.4911
acapbrsl(2,3)=2.1092
acapbrsl(3,1)=-1.4914
acapbrsl(3,2)=2.1092
acapbrsl(3,3)=2.7733
arbrsl(1,1)=-2.2823
arbrsl(1,2)=-2.5331
arbrsl(1,3)=-1.4914
arbrsl(1,4)=-2.5331
arbrsl(1,5)=-0.4911
arbrsl(1,6)=2.1092
arbrsl(1,7)=-1.4914
arbrsl(1,8)=2.1092
arbrsl(1,9)=2.7733
acbrsl(1,1)=-2.2823
acbrsl(2,1)=-2.5331
acbrsl(3,1)=-1.4914
acbrsl(4,1)=-2.5331
acbrsl(5,1)=-0.4911
acbrsl(6,1)=2.1092
acbrsl(7,1)=-1.4914
acbrsl(8,1)=2.1092
acbrsl(9,1)=2.7733
acapbrsd(1,1)=-2.2823
acapbrsd(1,2)=-2.5331
acapbrsd(1,3)=1.4914
acapbrsd(2,1)=-2.5331
acapbrsd(2,2)=-0.4911
acapbrsd(2,3)=-2.1092
acapbrsd(3,1)=1.4914
acapbrsd(3,2)=-2.1092
acapbrsd(3,3)=2.7733
```

```
arbrsd(1,1)=-2.2823
arbrsd(1,2)=-2.5331
arbrsd(1,3)=1.4914
arbrsd(1,4)=-2.5331
arbrsd(1,5)=-0.4911
arbrsd(1,6)=-2.1092
arbrsd(1,7)=1.4914
arbrsd(1,8)=-2.1092
arbrsd(1,9)=2.7733
acbrsd(1,1)=-2.2823
acbrsd(2,1)=-2.5331
acbrsd(3,1)=1.4914
acbrsd(4,1)=-2.5331
acbrsd(5,1)=-0.4911
acbrsd(6,1)=-2.1092
acbrsd(7,1)=1.4914
acbrsd(8,1)=-2.1092
acbrsd(9,1)=2.773
```

The above statements assign values to the polarizability tensors of *d* and *l* attached phenyl and *p*-Br-phenyl rings

```
call foel(1,3,1,3,e3,0.0,3,3)
call foel(1,9,1,9,e9,0.0,9,9)
e3(1,1)=1.0
e3(2,2)=1.0
e3(3,3)=1.0
e9(1,1)=1.0
e9(2,2)=1.0
e9(3,3)=1.0
e9(4,4)=1.0
e9(5,5)=1.0
e9(6,6)=1.0
e9(7,7)=1.0
e9(8,8)=1.0
e9(9,9)=1.0
call foel(1,1,1,22,j,0.0,1,22)
j(1,1)=1.0
call foel(1,1,1,52,jj,0.0,1,52)
jj(1,1)=1.0
call foel(1,22,1,1,js,0.0,22,1)
js(21,1)=1.0
js(22,1)=1.0
call foel(1,52,1,1,jjs,0.0,52,1)
jjs(51,1)=1.0
jjs(52,1)=1.0
call foel(1,6,1,6,tbarsdd,0.0,6,6)
call foel(1,18,1,18,txtbarsdd,0.0,18,18)
```

```
call foel(1,6,1,6,tbarsld,0.0,6,6)
call foel(1,18,1,18,txtbarsld,0.0,18,18)
call foel(1,6,1,6,tbarsdl,0.0,6,6)
call foel(1,18,1,18,txtbarsdl,0.0,18,18)
call foel(1,6,1,6,tbarsll,0.0,6,6)
call foel(1,18,1,18,txtbarsll,0.0,18,18)
call foel(1,6,1,6,tbarsd,0.0,6,6)
call foel(1,18,1,18,txtbarsd,0.0,18,18)
call foel(1,6,1,6,tbarsl,0.0,6,6)
call foel(1,18,1,18,txtbarsl,0.0,18,18)
.......................................................................................
```

The above statements fill all these matrices with 0s.

```
ph1=350.0
s1=sin(ph1/57.29578)
c1=cos(ph1/57.29578)
t1(2,1)=0.92718*c1
t1(2,2)=-0.37461*c1
t1(2,3)=s1
t1(3,1)=0.92718*s1
t1(3,2)=-0.37461*s1
t1(3,3)=-c1
call fill(1,1,3,3,t1,tbarsdd,3,3,6,6)
call fill(1,1,3,3,t1,tbarsld,3,3,6,6)
call dp(3,3,3,3,t1,t1,txt,9,9)
call fill(1,1,9,9,txt,txtbarsdd,9,9,18,18)
call fill(1,1,9,9,txt,txtbarsld,9,9,18,18)
ph1=250.0
s1=sin(ph1/57.29578)
c1=cos(ph1/57.29578)
t1(2,1)=0.92718*c1
t1(2,2)=-0.37461*c1
t1(2,3)=s1
t1(3,1)=0.92718*s1
t1(3,2)=-0.37461*s1
t1(3,3)=-c1
call fill(4,4,3,3,t1,tbarsdd,3,3,6,6)
call fill(4,4,3,3,t1,tbarsld,3,3,6,6)
call dp(3,3,3,3,t1,t1,txt,9,9)
call fill(10,10,9,9,txt,txtbarsdd,9,9,18,18)
call fill(10,10,9,9,txt,txtbarsld,9,9,18,18)
ph1=10.0
s1=sin(ph1/57.29578)
c1=cos(ph1/57.29578)
t1(2,1)=0.92718*c1
t1(2,2)=-0.37461*c1
t1(2,3)=s1
```

```
t1(3,1)=0.92718*s1
t1(3,2)=-0.37461*s1
t1(3,3)=-c1
call fill(1,1,3,3,t1,tbarsll,3,3,6,6)
call fill(1,1,3,3,t1,tbarsdl,3,3,6,6)
call dp(3,3,3,3,t1,t1,txt,9,9)
call fill(1,1,9,9,txt,txtbarsll,9,9,18,18)
call fill(1,1,9,9,txt,txtbarsdl,9,9,18,18)
ph1=110.0
s1=sin(ph1/57.29578)
c1=cos(ph1/57.29578)
t1(2,1)=0.92718*c1
t1(2,2)=-0.37461*c1
t1(2,3)=s1
t1(3,1)=0.92718*s1
t1(3,2)=-0.37461*s1
t1(3,3)=-c1
call fill(4,4,3,3,t1,tbarsll,3,3,6,6)
call fill(4,4,3,3,t1,tbarsdl,3,3,6,6)
call dp(3,3,3,3,t1,t1,txt,9,9)
call fill(10,10,9,9,txt,txtbarsll,9,9,18,18)
call fill(10,10,9,9,txt,txtbarsdl,9,9,18,18)
ph2=350.0
s2=sin(ph2/57.29578)
c2=cos(ph2/57.29578)
t2(2,1)=0.91355*c2
t2(2,2)=-0.40674*c2
t2(2,3)=s2
t2(3,1)=0.91355*s2
t2(3,2)=-0.40674*s2
t2(3,3)=-c2
call fill(1,1,3,3,t2,tbarsl,3,3,6,6)
call dp(3,3,3,3,t2,t2,txt,9,9)
call fill(1,1,9,9,txt,txtbarsl,9,9,18,18)
ph2=250.0
s2=sin(ph2/57.29578)
c2=cos(ph2/57.29578)
t2(2,1)=0.91355*c2
t2(2,2)=-0.40674*c2
t2(2,3)=s2
t2(3,1)=0.91355*s2
t2(3,2)=-0.40674*s2
t2(3,3)=-c2
call fill(4,4,3,3,t2,tbarsl,3,3,6,6)
call dp(3,3,3,3,t2,t2,txt,9,9)
call fill(10,10,9,9,txt,txtbarsl,9,9,18,18)
ph2=10.0
```

```
s2=sin(ph2/57.29578)
c2=cos(ph2/57.29578)
t2(2,1)=0.91355*c2
t2(2,2)=-0.40674*c2
t2(2,3)=s2
t2(3,1)=0.91355*s2
t2(3,2)=-0.40674*s2
t2(3,3)=-c2
call fill(1,1,3,3,t2,tbarsd,3,3,6,6)
call dp(3,3,3,3,t2,t2,txt,9,9)
call fill(1,1,9,9,txt,txtbarsd,9,9,18,18)
ph2=110.0
s2=sin(ph2/57.29578)
c2=cos(ph2/57.29578)
t2(2,1)=0.91355*c2
t2(2,2)=-0.40674*c2
t2(2,3)=s2
t2(3,1)=0.91355*s2
t2(3,2)=-0.40674*s2
t2(3,3)=-c2
call fill(4,4,3,3,t2,tbarsd,3,3,6,6)
call dp(3,3,3,3,t2,t2,txt,9,9)
call fill(10,10,9,9,txt,txtbarsd,9,9,18,18)
call foel(1,22,1,22,asl,0.0,22,22)
call foel(1,22,1,22,asd,0.0,22,22)
call foel(1,22,1,22,asll,0.0,22,22)
call foel(1,22,1,22,asdl,0.0,22,22)
call foel(1,22,1,22,asdd,0.0,22,22)
call foel(1,22,1,22,asld,0.0,22,22)
call foel(1,52,1,52,qsl,0.0,52,52)
call foel(1,52,1,52,qsd,0.0,52,52)
call foel(1,52,1,52,qsll,0.0,52,52)
call foel(1,52,1,52,qsdl,0.0,52,52)
call foel(1,52,1,52,qsdd,0.0,52,52)
call foel(1,52,1,52,qsld,0.0,52,52)
call foel(1,22,1,22,abrsl,0.0,22,22)
call foel(1,22,1,22,abrsd,0.0,22,22)
call foel(1,52,1,52,qbrsl,0.0,52,52)
call foel(1,52,1,52,qbrsd,0.0,52,52)
call fill(1,1,2,2,up,asl,2,2,22,22)
call fill(1,1,2,2,up,asd,2,2,22,22)
call fill(1,1,2,2,up,abrsl,2,2,22,22)
call fill(1,1,2,2,up,abrsd,2,2,22,22)
call fill(1,1,2,2,uppm,asll,2,2,22,22)
call fill(1,1,2,2,uppm,asdd,2,2,22,22)
call fill(1,1,2,2,uppr,asld,2,2,22,22)
call fill(1,1,2,2,uppr,asdl,2,2,22,22)
```

```
call fill(21,21,2,2,up,asl,2,2,22,22)
call fill(21,21,2,2,up,asd,2,2,22,22)
call fill(21,21,2,2,up,abrsl,2,2,22,22)
call fill(21,21,2,2,up,abrsd,2,2,22,22)
call fill(21,21,2,2,uppm,asll,2,2,22,22)
call fill(21,21,2,2,uppm,asdd,2,2,22,22)
call fill(21,21,2,2,uppr,asld,2,2,22,22)
call fill(21,21,2,2,uppr,asdl,2,2,22,22)
call fill(1,1,2,2,up,qsl,2,2,52,52)
call fill(1,1,2,2,up,qsd,2,2,52,52)
call fill(1,1,2,2,uppm,qsll,2,2,52,52)
call fill(1,1,2,2,uppm,qsdd,2,2,52,52)
call fill(1,1,2,2,uppr,qsdl,2,2,52,52)
call fill(1,1,2,2,uppr,qsld,2,2,52,52)
call fill(1,1,2,2,up,qbrsl,2,2,52,52)
call fill(1,1,2,2,up,qbrsd,2,2,52,52)
call fill(51,51,2,2,up,qsl,2,2,52,52)
call fill(51,51,2,2,up,qsd,2,2,52,52)
call fill(51,51,2,2,uppm,qsll,2,2,52,52)
call fill(51,51,2,2,uppm,qsdd,2,2,52,52)
call fill(51,51,2,2,uppr,qsdl,2,2,52,52)
call fill(51,51,2,2,uppr,qsld,2,2,52,52)
call fill(51,51,2,2,up,qbrsl,2,2,52,52)
call fill(51,51,2,2,up,qbrsd,2,2,52,52)
call dp(2,2,9,9,up,e9,uxe9,18,18)
call mp(18,18,18,uxe9,txtbarsl,uett)
call fill(3,3,18,18,uett,asl,18,18,22,22)
call fill(3,3,18,18,uett,abrsl,18,18,22,22)
call mp(18,18,18,uxe9,txtbarsd,uett)
call fill(3,3,18,18,uett,asd,18,18,22,22)
call fill(3,3,18,18,uett,abrsd,18,18,22,22)
call dp(2,2,9,9,uppm,e9,uxe9,18,18)
call mp(18,18,18,uxe9,txtbarsll,uett)
call fill(3,3,18,18,uett,asll,18,18,22,22)
call mp(18,18,18,uxe9,txtbarsdd,uett)
call fill(3,3,18,18,uett,asdd,18,18,22,22)
call dp(2,2,9,9,uppr,e9,uxe9,18,18)
call mp(18,18,18,uxe9,txtbarsld,uett)
call fill(3,3,18,18,uett,asld,18,18,22,22)
call mp(18,18,18,uxe9,txtbarsdl, uett)
call fill(3,3,18,18,uett,asdl,18,18,22,22)
call dp(2,2,1,9,up,arsl,uxa,2,18)
call mp(2,18,18,uxa,txtbarsl,uatt)
call fill(1,3,2,18,uatt,asl,2,18,22,22)
call fill(1,9,2,18,uatt,qsl,2,18,52,52)
call dp(2,2,1,9,up,arbrsl,uxa,2,18)
call mp(2,18,18,uxa,txtbarsl,uatt)
```

```
      call fill(1,3,2,18,uatt,abrsl,2,18,22,22)
      call fill(1,9,2,18,uatt,qbrsl,2,18,52,52)
      call dp(2,2,1,9,up,arsd,uxa,2,18)
      call mp(2,18,18,uxa,txtbarsd,uatt)
      call fill(1,3,2,18,uatt,asd,2,18,22,22)
      call fill(1,9,2,18,uatt,qsd,2,18,52,52)
      call dp(2,2,1,9,up,arbrsd,uxa,2,18)
      call mp(2,18,18,uxa,txtbarsd, uatt)
      call fill(1,3,2,18,uatt,abrsd,2,18,22,22)
      call fill(1,9,2,18,uatt,qbrsd,2,18,52,52)
      call mp(1,9,1,arsl,acsl,arac)
      do 1 ij=1,2
      do 1 ji=1,2
    1 aau(ij,ji)=arac*up(ij, ji)
      call fill(1,21,2,2,aau,asl,2,2,22,22)
      call mp(1,9,1,arsd,acsd,arac)
      do 2 ij=1,2
      do 2 ji=1,2
    2 aau(ij,ji)=arac*up(ij,ji)
      call fill(1,21,2,2,aau,asd,2,2,22,22)
      call mp(1,9,1,arbrsl,acbrsl,arac)
      do 3 ij=1,2
      do 3 ji=1,2
    3 aau(ij,ji)=arac*up(ij,ji)
      call fill(1,21,2,2,aau,abrsl,2,2,22,22)
      call mp(1,9,1,arbrsd,acbrsd,arac)
      do 4 ij=1,2
      do 4 ji=1,2
    4 aau(ij,ji)=arac*up(ij,ji)
      call fill(1,21,2,2,aau,abrsd,2,2,22,22)
      call dp(2,2,9,1,up,acsl,uxac,18,2)
      call fill(3,21,18,2,uxac,asl,18,2,22,22)
      call dp(2,2,9,1,up,acsd,uxac,18,2)
      call fill(3,21,18,2,uxac,asd,18,2,22,22)
      call dp(2,2,9,1,up,acbrsl,uxac,18,2)
      call fill(3,21,18,2,uxac,abrsl,18,2,22,22)
      call dp(2,2,9,1,up,acbrsd,uxac,18,2)
      call fill(3,21,18,2,uxac,abrsd,18,2,22,22)
      call dp(2,2,3,3,up,e3,uxe3,6,6)
      call mp(6,6,6,uxe3,tbarsl,uet)
      call fill(3,3,6,6,uet,qsl,6,6,52,52)
      call fill(45,45,6,6,uet,qsl,6,6,52,52)
      call fill(3,3,6,6,uet,qbrsl,6,6,52,52)
      call fill(45,45,6,6,uet,qbrsl,6,6,52,52)
      call mp(6,6,6,uxe3,tbarsd,uet)
      call fill(3,3,6,6,uet,qsd,6,6,52,52)
```

```
call fill(45,45,6,6,uet,qsd,6,6,52,52)
call fill(3,3,6,6,uet,qbrsd,6,6,52,52)
call fill(45,45,6,6,uet,qbrsd,6,6,52,52)
call dp(2,2,3,3,uppm,e3,uxe3,6,6)
call mp(6,6,6,uxe3,tbarsll,uet)
call fill(3,3,6,6,uet,qsll,6,6,52,52)
call fill(45,45,6,6,uet,qsll,6,6,52,52)
call mp(6,6,6,uxe3,tbarsdd,uet)
call fill(3,3,6,6,uet,qsdd,6,6,52,52)
call fill(45,45,6,6,uet,qsdd,6,6,52,52)
call dp(2,2,3,3,uppr,e3,uxe3,6,6)
call mp(6,6,6,uxe3,tbarsld,uet)
call fill(3,3,6,6,uet,qsld,6,6,52,52)
call fill(45,45,6,6,uet,qsld,6,6,52,52)
call mp(6,6,6,uxe3,tbarsdl,uet)
call fill(3,3,6,6,uet,qsdl,6,6,52,52)
call fill(45,45,6,6,uet,qsdl,6,6,52,52)
call dp(2,2,9,9,up,e9,uxe9,18,18)
call mp(18,18,18,uxe9,txtbarsl,uett)
call fill(9,9,18,18,uett,qsl,18,18,52,52)
call fill(27,27,18,18,uett,qsl,18,18,52,52)
call fill(9,9,18,18,uett,qbrsl,18,18,52,52)
call fill(27,27,18,18,uett,qbrsl,18,18,52,52)
call mp(18,18,18,uxe9,txtbarsd,uett)
call fill(9,9,18,18,uett,qsd,18,18,52,52)
call fill(27,27,18,18,uett,qsd,18,18,52,52)
call fill(9,9,18,18,uett,qbrsd,18,18,52,52)
call fill(27,27,18,18,uett,qbrsd,18,18,52,52)
call dp(2,2,9,9,uppm,e9,uxe9,18,18)
call mp(18,18,18,uxe9,txtbarsll,uett)
call fill(9,9,18,18,uett,qsll,18,18,52,52)
call fill(27,27,18,18,uett,qsll,18,18,52,52)
call mp(18,18,18,uxe9,txtbarsdd,uett)
call fill(9,9,18,18,uett,qsdd,18,18,52,52)
call fill(27,27,18,18,uett,qsdd,18,18,52,52)
call dp(2,2,9,9,uppr,e9,uxe9,18,18)
call mp(18,18,18,uxe9,txtbarsld,uett)
call fill(9,9,18,18,uett,qsld,18,18,52,52)
call fill(27,27,18,18,uett,qsld,18,18,52,52)
call mp(18,18,18,uxe9,txtbarsdl,uett)
call fill(9,9,18,18,uett,qsdl,18,18,52,52)
call fill(27,27,18,18,uett,qsdl,18,18,52,52)
call dp(1,3,1,3,mutsl,mutsl,muxmu,1,9)
call dp(2,2,1,9,up,muxmu,uxmuxmu,2,18)
do 5 ij=1,2
do 5 ji=1,18
```

```
5 umumu(ij, ji)=0.5*uxmuxmu(ij,ji)
  call mp(2,18,18,umumu,txtbarsl,uuutt)
  call fill(1,27,2,18,uuutt,qsl,2,18,52,52)
  call dp(1,3,1,3,mutsd,mutsd,muxmu,1,9)
  call dp(2,2,1,9,up,muxmu,uxmuxmu,2,18)
  do 6 ij=1,2
  do 6 ji=1,18
6 umumu(ij, ji)=0.5*uxmuxmu(ij,ji)
  call mp(2,18,18,umumu,txtbarsd,uuutt)
  call fill(1,27,2,18,uuutt,qsl,2,18,52,52)
  call dp(1,3,1,3,mutbrsl,mutbrsl,muxmu,1,9)
  call dp(2,2,1,9,up,muxmu,uxmuxmu,2,18)
  do 7 ij=1,2
  do 7 ji=1,18
7 umumu(ij,ji)=0.5*uxmuxmu(ij,ji)
  call mp(2,18,18,umumu,txtbarsl,uuutt)
  call fill(1,27,2,18,uuutt,qbrsl,2,18,52,52)
  call dp(1,3,1,3,mutbrsd,mutbrsd,muxmu,1,9)
  call dp(2,2,1,9,up,muxmu,uxmuxmu,2,18)
  do 8 ij=1,2
  do 8 ji=1,18
8 umumu(ij,ji)=0.5*uxmuxmu(ij,ji)
  call mp(2,18,18,umumu,txtbarsd,uuutt)
  call fill(1,27,2,18,uuutt,qsl,2,18,52,52)
  call dp(3,1,3,3,musl,e3,muxe,9,3)
  call mp(1,9,3,arsl,muxe,amue)
  call dp(2,2,1,3,up,amue,uamue,2,6)
  call mp(2,6,6,uamue,tbarsl,uamuet)
  call fill(1,45,2,6,uamuet,qsl,2,6,52,52)
  call dp(3,1,3,3,musd,e3,muxe,9,3)
  call mp(1,9,3,arsd,muxe,amue)
  call dp(2,2,1,3,up,amue,uamue,2,6)
  call mp(2,6,6,uamue,tbarsd, uamuet)
  call fill(1,45,2,6,uamuet,qsd,2,6,52,52)
  call dp(3,1,3,3,mubrsl,e3,muxe,9,3)
  call mp(1,9,3,arbrsl,muxe,amue)
  call dp(2,2,1,3,up,amue,uamue,2,6)
  call mp(2,6,6,uamue,tbarsl,uamuet)
  call fill(1,45,2,6,uamuet,qbrsl,2,6,52,52)
  call dp(3,1,3,3,mubrsd,e3,muxe,9,3)
  call mp(1,9,3,arbrsd,muxe,amue)
  call dp(2,2,1,3,up,amue,uamue,2,6)
  call mp(2,6,6,uamue,tbarsd,uamuet)
  call fill(1,45,2,6,uamuet,qbrsd,2,6,52,52)
  call dp(3,1,3,1,musl,musl,mumu,9,1)
  call mp(1,9,1,arsl,mumu,armumu)
```

```
      do 9 ij=1,2
      do 9 ji=1,2
    9 uamumu(ij, ji)=0.5*armumu*up(ij, ji)
      call fill(1,51,2,2,uamumu,qsl,2,2,52,52)
      call dp(3,1,3,1,musd,musd,mumu,9,1)
      call mp(1,9,1,arsd,mumu,armumu)
      do 10 ij=1,2
      do 10 ji=1,2
   10 uamumu(ij, ji)=0.5*armumu*up(ij, ji)
      call fill(1,51,2,2,uamumu,qsd,2,2,52,52)
      call dp(3,1,3,1,mubrsl,mubrsl,mumu,9,1)
      call mp(1,9,1,arbrsl,mumu,armumu)
      do 11 ij=1,2
      do 11 ji=1,2
   11 uamumu(ij, ji)=0.5*armumu*up(ij, ji)
      call fill(1,51,2,2,uamumu,qbrsl,2,2,52,52)
      call dp(3,1,3,1,mubrsd,mubrsd,mumu,9,1)
      call mp(1,9,1,arbrsd,mumu,armumu)
      do 12 ij=1,2
      do 12 ji=1,2
   12 uamumu(ij,ji)=0.5*armumu*up(ij,ji)
      call fill(1,51,2,2,uamumu,qbrsd,2,2,52,52)
      call dp(3,3,1,3,e3,mutsl,exmut,3,9)
      call dp(2,2,3,9,up, exmut,uexmut,6,18)
      call mp(6,18,18,uexmut,txtbarsl,uemutt)
      call fill(3,27,6,18,uemutt,qsl,6,18,52,52)
      call dp(3,3,1,3,e3,mutsd,exmut,3,9)
      call dp(2,2,3,9,up,exmut,uexmut,6,18)
      call mp(6,18,18,uexmut,txtbarsd,uemutt)
      call fill(3,27,6,18,uemutt,qsd,6,18,52,52)
      call dp(3,3,1,3,e3,mutbrsl,exmut,3,9)
      call dp(2,2,3,9,up,exmut,uexmut,6,18)
      call mp(6,18,18,uexmut,txtbarsl,uemutt)
      call fill(3,27,6,18,uemutt,qbrsl,6,18,52,52)
      call dp(3,3,1,3,e3,mutbrsd, exmut,3,9)
      call dp(2,2,3,9,up, exmut, uexmut,6,18)
      call mp(6,18,18,uexmut,txtbarsd,uemutt)
      call fill(3,27,6,18,uemutt,qbrsd,6,18,52,52)
      call dp(2,2,3,3,up,acapsl,uacap,6,6)
      call mp(6,6,6,uacap,tbarsl,uat)
      call fill(3,45,6,6,uat,qsl,6,6,52,52)
      call dp(2,2,3,3,up,acapsd,uacap,6,6)
      call mp(6,6,6,uacap,tbarsd,uat)
      call fill(3,45,6,6,uat,qsd,6,6,52,52)
      call dp(2,2,3,3,up,acapbrsl,uacap,6,6)
      call mp(6,6,6,uacap,tbarsl,uat)
```

```
      call fill(3,45,6,6,uat, qbrsl,6,6,52,52)
      call dp(2,2,3,3,up,acapbrsd,uacap,6,6)
      call mp(6,6,6,uacap,tbarsd,uat)
      call fill(3,45,6,6,uat,qbrsd,6,6,52,52)
      call dp(3,3,1,3,e3,mutsl,exmut,3,9)
      call mp(3,9,1,exmut,acsl,emuac)
      call dp(2,2,3,1,up,emuac,uemuac,6,2)
      call fill(3,51,6,2,uemuac,qsl,6,2,52,52)
      call dp(3,3,1,3,e3,mutsd,exmut,3,9)
      call mp(3,9,1,exmut,acsd,emuac)
      call dp(2,2,3,1,up,emuac,uemuac,6,2)
      call fill(3,51,6,2,uemuac,qsd,6,2,52,52)
      call dp(3,3,1,3,e3,mutbrsl,exmut,3,9)
      call mp(3,9,1,exmut,acbrsl,emuac)
      call dp(2,2,3,1,up,emuac,uemuac,6,2)
      call fill(3,51,6,2,uemuac,qbrsl,6,2,52,52)
      call dp(3,3,1,3,e3,mutbrsd,exmut,3,9)
      call mp(3,9,1,exmut,acbrsd,emuac)
      call dp(2,2,3,1,up,emuac,uemuac,6,2)
      call fill(3,51,6,2,uemuac,qbrsd,6,2,52,52)
      call dp(3,1,3,3,musl,e3,muxe,9,3)
      call dp(2,2,9,3,up,muxe,umuxe,18,6)
      call mp(18,6,6,umuxe,tbarsl,umuxet)
      call fill(9,45,18,6,umuxet,qsl,18,6,52,52)
      call dp(3,1,3,3,musd,e3,muxe,9,3)
      call dp(2,2,9,3,up,muxe,umuxe,18,6)
      call mp(18,6,6,umuxe,tbarsd,umuxet)
      call fill(9,45,18,6,umuxet,qsd,18,6,52,52)
      call dp(3,1,3,3,mubrsl,e3,muxe,9,3)
      call dp(2,2,9,3,up,muxe,umuxe,18,6)
      call mp(18,6,6,umuxe,tbarsl,umuxet)
      call fill(9,45,18,6,umuxet,qbrsl,18,6,52,52)
      call dp(3,1,3,3,mubrsd,e3,muxe,9,3)
      call dp(2,2,9,3,up,muxe,umuxe,18,6)
      call mp(18,6,6,umuxe,tbarsd,umuxet)
      call fill(9,45,18,6,umuxet,qbrsd,18,6,52,52)
      call dp(3,1,3,1,musl,musl,mumu,9,1)
      call dp(2,2,9,1,up,mumu,umumu,18,2)
      do 13 ij=1,18
      do 13 ji=1,2
13    mumumu(ij,ji)=0.5*umumu(ij,ji)
      call fill(9,51,18,2,mumumu,qsl,18,2,52,52)
      call dp(3,1,3,1,musd,musd,mumu,9,1)
      call dp(2,2,9,1,up,mumu,umumu,18,2)
      do 14 ij=1,18
      do 14 ji=1,2
```

```
14 mumumu(ij,ji)=0.5*umumu(ij,ji)
   call fill(9,51,18,2,mumumu,qsd,18,2,52,52)
   call dp(3,1,3,1,mubrsl,mubrsl,mumu,9,1)
   call dp(2,2,9,1,up,mumu,umumu,18,2)
   do 15 ij=1,18
   do 15 ji=1,2
15 mumumu(ij,ji)=0.5*umumu(ij,ji)
   call fill(9,51,18,2,mumumu,qbrsl,18,2,52,52)
   call dp(3,1,3,1,mubrsd,mubrsd,mumu,9,1)
   call dp(2,2,9,1,up,mumu,umumu,18,2)
   do 16 ij=1,18
   do 16 ji=1,2
16 mumumu(ij,ji)=0.5*umumu(ij,ji)
   call fill(9,51,18,2,mumumu,qbrsd,18,2,52,52)
   call dp(2,2,9,1,up,acsl,uxac,18,2)
   call fill(27,51,18,2,uxac,qsl,18,2,52,52)
   call dp(2,2,9,1,up,acsd,uxac,18,2)
   call fill(27,51,18,2,uxac,qsd,18,2,52,52)
   call dp(2,2,9,1,up,acbrsl,uxac,18,2)
   call fill(27,51,18,2,uxac,qbrsl,18,2,52,52)
   call dp(2,2,9,1,up,acbrsd,uxac,18,2)
   call fill(27,51,18,2,uxac,qbrsd,18,2,52,52)
   call dp(2,2,3,1,up,musl,uxmu,6,2)
   call fill(45,51,6,2,uxmu,qsl,6,2,52,52)
   call dp(2,2,3,1,up,musd,uxmu,6,2)
   call fill(45,51,6,2,uxmu,qsd,6,2,52,52)
   call dp(2,2,3,1,up,mubrsl,uxmu,6,2)
   call fill(45,51,6,2,uxmu,qbrsl,6,2,52,52)
   call dp(2,2,3,1,up,mubrsd,uxmu,6,2)
   call fill(45,51,6,2,uxmu,qbrsd,6,2,52,52)
```

The above statements create the elements for and fill the $q = \mu^T\alpha\mu$ and a = $\alpha^R\acute{\alpha}^C$ matrices needed for evaluation of ${}_mK$ and of course are different for d and *l* S and *p*-BrS repeat units.

```
   pr=0.50
   do 6969 kij=1,4
   pr=pr+0.01
   dum=1
   ketot=0.0
   kaatot=0.0
   kuautot=0.0
   do 69 ijk=1,200
   pbrs=1.0
   numbrs=0.0
   dorl=1.0
   dum=dum+1
```

```
   call random(dum,rn)
   if (rn.le. pr) go to 33
   call mp(2,2,2,up,uppm,u1)
   dum=dum+1
   call random(dum,rn)
   if (rn.lt. pbrs) go to 333
   call mp(1,22,22,j,asd,ja)
   call mp(1,22,22,ja,asdd,jaa)
   call mp(1,52,52,jj,qsd,jjq)
   call mp(1,52,52,jjq,qsdd,jjqq)
   dorl=1.0
   go to 43
333 call mp(1,22,22,j,abrsd,ja)
   call mp(1,22,22,ja,asdd,jaa)
   call mp(1,52,52,jj,qbrsd,jjq)
   call mp(1,52,52,jjq,qsdd,jjqq)
   numbrs=numbrs+1.0
   dorl=1.0
   go to 43
33 call mp(2,2,2,up,uppr,u1)
   dum=dum+1
   call random(dum, rn)
   if (rn.lt. pbrs) go to 3333
   call mp(1,22,22,j,asd,ja)
   call mp(1,22,22,ja,asdl,jaa)
   call mp(1,52,52,jj,qsd,jjq)
   call mp(1,52,52,jjq,qsdl,jjqq)
   dorl=0.0
   go to 43
3333 call mp(1,22,22,j,abrsd,ja)
   call mp(1,22,22,ja,asdl,jaa)
   call mp(1,52,52,jj,qbrsd,jjq)
   call mp(1,52,52,jjq,qsdl,jjqq)
   numbrs=numbrs+1.0
   dorl=0.0
43 do 96 jki=1,89
   pbrs=1.0
   dum=dum+1
   call random(dum,rn)
   if (rn.le. pr) go to 98
   if (dorl.gt. 0.5) go to 89
   call mp(2,2,2,u1,up,u2)
   call mp(2,2,2,u2,uppm,u1)
   dum=dum+1
   call random(dum,rn)
   if (rn.lt. pbrs) go to 189
   call mp(1,22,22,jaa,asl,ja)
```

```
   call mp(1,52,52,jjqq,qsl,jjq)
   call mp(1,22,22,ja,asll,jaa)
   call mp(1,52,52,jjq,qsll,jjqq)
   dorl=0.0
   go to 96
189 call mp(1,22,22,jaa,abrsl,ja)
   call mp(1,52,52,jjqq,qbrsl,jjq)
   call mp(1,22,22,ja,asll,jaa)
   call mp(1,52,52,jjq,qsll,jjqq)
   numbrs=numbrs+1.0
   dorl=0.0
   go to 96
89 call mp(2,2,2,u1,up,u2)
   call mp(2,2,2,u2,uppm,u1)
   dum=dum+1
   call random(dum,rn)
   if (rn.lt. pbrs) go to 1189
   call mp(1,22,22,jaa,asd,ja)
   call mp(1,52,52,jjqq,qsd,jjq)
   call mp(1,22,22,ja,asdd,jaa)
   call mp(1,52,52,jjq,qsdd,jjqq)
   dorl=1.0
   go to 96
1189 call mp(1,22,22,jaa,abrsd,ja)
   call mp(1,52,52,jjqq,qbrsd,jjq)
   call mp(1,22,22,ja,asdd,jaa)
   call mp(1,52,52,jjq,qsdd,jjqq)
   numbrs=numbrs+1.0
   dorl=1.0
   go to 96
98 if (dorl.gt. 0.5) go to 79
   call mp(2,2,2,u1,up,u2)
   call mp(2,2,2,u2,uppr,u1)
   dum=dum+1
   call random(dum,rn)
   if (rn.lt. pbrs) go to 2189
   call mp(1,22,22,jaa,asl,ja)
   call mp(1,52,52,jjqq,qsl,jjq)
   call mp(1,22,22,ja,asld,jaa)
   call mp(1,52,52,jjq,qsld,jjqq)
   dorl=1.0
   go to 96
2189 call mp(1,22,22,jaa,abrsl,ja)
   call mp(1,52,52,jjqq,qbrsl,jjq)
   call mp(1,22,22,ja,asld,jaa)
   call mp(1,52,52,jjq,qsld,jjqq)
   numbrs=numbrs+1.0
```

```
   dorl=1.0
   go to 96
79 call mp(2,2,2,u1,up,u2)
   call mp(2,2,2,u2,uppr,u1)
   dum=dum+1
   call random(dum,rn)
   if (rn.lt. pbrs) go to 3189
   call mp(1,22,22,jaa,asd,ja)
   call mp(1,52,52,jjqq,qsd,jjq)
   call mp(1,22,22,ja,asdl,jaa)
   call mp(1,52,52,jjq,qsdl,jjqq)
   dorl=0.0
   go to 96
3189 call mp(1,22,22,jaa,abrsd,ja)
   call mp(1,52,52,jjqq,qbrsd,jjq)
   call mp(1,22,22,ja,asdl,jaa)
   call mp(1,52,52,jjq,qsdl,jjqq)
   numbrs=numbrs+1.0
   dorl=0.0
96 continue
   do 966 jki=1,120
   pbrs=0.0
   dum=dum+1
   call random(dum,rn)
   if (rn.le. pr) go to 988
   if (dorl.gt. 0.5) go to 899
   call mp(2,2,2,u1,up,u2)
   call mp(2,2,2,u2,uppm,u1)
   dum=dum+1
   call random(dum,rn)
   if (rn.lt. pbrs) go to 1899
   call mp(1,22,22,jaa,asl,ja)
   call mp(1,52,52,jjqq,qsl,jjq)
   call mp(1,22,22,ja,asll,jaa)
   call mp(1,52,52,jjq,qsll,jjqq)
   dorl=0.0
   go to 966
1899 call mp(1,22,22,jaa,abrsl,ja)
   call mp(1,52,52,jjqq,qbrsl,jjq)
   call mp(1,22,22,ja,asll,jaa)
   call mp(1,52,52,jjq,qsll,jjqq)
   numbrs=numbrs+1.0
   dorl=0.0
   go to 966
899 call mp(2,2,2,u1,up,u2)
   call mp(2,2,2,u2,uppm,u1)
   dum=dum+1
```

```
   call random(dum, rn)
   if (rn.lt. pbrs) go to 11899
   call mp(1,22,22,jaa,asd,ja)
   call mp(1,52,52,jjqq,qsd,jjq)
   call mp(1,22,22,ja,asdd,jaa)
   call mp(1,52,52,jjq,qsdd,jjqq)
   dorl=1.0
   go to 966
11899 call mp(1,22,22,jaa,abrsd,ja)
   call mp(1,52,52,jjqq,qbrsd,jjq)
   call mp(1,22,22,ja,asdd,jaa)
   call mp(1,52,52,jjq,qsdd,jjqq)
   numbrs=numbrs+1.0
   dorl=1.0
   go to 966
988 if (dorl. gt. 0.5) go to 799
   call mp(2,2,2,u1,up,u2)
   call mp(2,2,2,u2,uppr,u1)
   dum=dum+1
   call random(dum,rn)
   if (rn.lt. pbrs) go to 21899
   call mp(1,22,22,jaa,asl,ja)
   call mp(1,52,52,jjqq,qsl,jjq)
   call mp(1,22,22,ja,asld,jaa)
   call mp(1,52,52,jjq,qsld,jjqq)
   dorl=1.0
   go to 966 21899 call mp(1,22,22,jaa,abrsl,ja)
   call mp(1,52,52,jjqq,qbrsl,jjq)
   call mp(1,22,22,ja,asld,jaa)
   call mp(1,52,52,jjq,qsld,jjqq)
   numbrs=numbrs+1.0
   dorl=1.0
   go to 966
799 call mp(2,2,2,u1,up,u2)
   call mp(2,2,2,u2,uppr,u1)
   dum=dum+1
   call random(dum,rn)
   if (rn.lt. pbrs) go to 31899
   call mp(1,22,22,jaa,asd,ja)
   call mp(1,52,52,jjqq,qsd,jjq)
   call mp(1,22,22,ja,asdl,jaa)
   call mp(1,52,52,jjq,qsdl,jjqq)
   dorl=0.0
   go to 966
31899 call mp(1,22,22,jaa,abrsd,ja)
   call mp(1,52,52,jjqq,qbrsd,jjq)
   call mp(1,22,22,ja,asdl,jaa)
```

```
   call mp(1,52,52,jjq,qsdl,jjqq)
   numbrs=numbrs+1.0
   dorl=0.0
966 continue
   do 969 jki=1,90
   pbrs=1.0
   dum=dum+1
   call random(dum,rn)
   if (rn.le. pr) go to 989
   if (dorl.gt. 0.5) go to 898
   call mp(2,2,2,u1,up,u2)
   call mp(2,2,2,u2,uppm,u1)
   dum=dum+1
   call random(dum,rn)
   if (rn.lt. pbrs) go to 1891
   call mp(1,22,22,jaa,asl,ja)
   call mp(1,52,52,jjqq,qsl,jjq)
   call mp(1,22,22,ja,asll,jaa)
   call mp(1,52,52,jjq,qsll,jjqq)
   dorl=0.0
   go to 969
1891 call mp(1,22,22,jaa,abrsl,ja)
   call mp(1,52,52,jjqq,qbrsl,jjq)
   call mp(1,22,22,ja,asll,jaa)
   call mp(1,52,52,jjq,qsll,jjqq)
   numbrs=numbrs+1.0
   dorl=0.0
   go to 969
898 call mp(2,2,2,u1,up,u2)
   call mp(2,2,2,u2,uppm,u1)
   dum=dum+1
   call random(dum,rn)
   if (rn. lt. pbrs) go to 1188
   call mp(1,22,22,jaa,asd,ja)
   call mp(1,52,52,jjqq,qsd,jjq)
   call mp(1,22,22,ja,asdd,jaa)
   call mp(1,52,52,jjq,qsdd,jjqq)
   dorl=1.0
   go to 969
1188 call mp(1,22,22,jaa,abrsd,ja)
   call mp(1,52,52,jjqq,qbrsd,jjq)
   call mp(1,22,22,ja,asdd,jaa)
   call mp(1,52,52,jjq,qsdd,jjqq)
   numbrs=numbrs+1.0
   dorl=1.0
   go to 969
```

```
989 if (dorl .gt. 0.5) go to 797
   call mp(2,2,2,u1,up,u2)
   call mp(2,2,2,u2,uppr,u1)
   dum=dum+1
   call random(dum,rn)
   if (rn.lt. pbrs) go to 1892
   call mp(1,22,22,jaa,asl,ja)
   call mp(1,52,52,jjqq, qsl,jjq)
   call mp(1,22,22,ja,asld,jaa)
   call mp(1,52,52,jjq,qsld,jjqq)
   dorl=1.0
   go to 969
1892 call mp(1,22,22,jaa,abrsl,ja)
   call mp(1,52,52,jjqq, qbrsl,jjq)
   call mp(1,22,22,ja,asld,jaa)
   call mp(1,52,52,jjq,qsld,jjqq)
   numbrs=numbrs+1.0
   dorl=1.0
   go to 969
797 call mp(2,2,2,u1,up,u2)
   call mp(2,2,2,u2,uppr,u1)
   dum=dum+1
   call random(dum, rn)
   if (rn. lt. pbrs) go to 1893
   call mp(1,22,22,jaa,asd,ja)
   call mp(1,52,52,jjqq,qsd,jjq)
   call mp(1,22,22,ja,asdl,jaa)
   call mp(1,52,52,jjq,qsdl,jjqq)
   dorl=0.0
   go to 969
1893 call mp(1,22,22,jaa,abrsd,ja)
   call mp(1,52,52,jjqq,qbrsd,jjq)
   call mp(1,22,22,ja,asdl,jaa)
   call mp(1,52,52,jjq,qsdl,jjqq)
   numbrs=numbrs+1.0
   dorl=0.0
969 continue
   z=u1(1,1)+u1(1,2)
   print 201, numbrs
   print 111, z
   call mp(1,22,1,jaa,js,jaajs)
   keaa=(2.0*jaajs)/z
   call mp(1,52,1,jjqq,jjs,jjuaujjs)
   keuau=(2.0*jjuaujjs)/z
   ke=keaa+keuau
   print 100, keaa,keuau,ke
```

```
   ketot=ketot+ke
   kaatot=kaatot+keaa
   kuautot=kuautot+keuau
69 continue
   kaatot=kaatot/200.0
   kuautot=kuautot/200.0
   ketot=ketot/200.0
   print 201, pr
   print 100, kaatot,kuautot,ketot
6969 continue
```

The statements above create each atactic triblock chain one repeat unit at a time and multiply the appropriate *U*, *q*, and *a* matrices for both bonds in a repeat unit to their products for the preceding repeat units, so we may obtain Z, $<\mu^T\alpha\mu>$, and $<\alpha^R\acute{\alpha}^C>$ and, therefore:

$$_{m}\mathbf{K} = (2\pi\mathbf{N_A}/135)\left[\left(<\mu^T\alpha\mu>\right)/\mathbf{k^2T^2} + \left(<\alpha^R\acute{\alpha}^C>\right)/\mathbf{kT}\right].$$

This is then repeated for a total of 200 p-$BrS_{90}S_{120}$-p-BrS_{90} triblocks to obtain $\langle {}_m K \rangle$s.

```
   stop
   end
```

```
   subroutine foel(a,b,c,d,aa,val,d1,d2)
   integer a,b,c,d,ii,jj,d1,d2
   real val,aa(d1,d2)
   do 88 ii=a,b
   do 88 jj=c,d
88 aa(ii,jj)=val
   return
   end
   subroutine mp(r,s,t,a,b,c)
   integer r,s,t,ii,jj,kk
   real sum,a(r,s),b(s,t),c(r,t)
   do 55 ii=1,r
   do 55 jj=1,t
   sum=0
   do 77 kk=1,s
77 sum=a(ii,kk)*b(kk,jj)+sum
55 c(ii,jj)=sum
   return
   end
   subroutine dp(m,n,p,s,aa,bb,cc,pr1,pr2)
   integer m,n,p,s,ii,jj,kk,ll,pr1,pr2,a,b
   real aa(m,n),bb(p,s),cc(pr1,pr2)
   do 22 ii=1,m
```

```
   do 22 jj=1,n
   do 22 kk=1,p
   do 22 ll=1,s
   a=((ii-1)*p)+kk
   b=((jj-1)*s)+ll
22 cc(a,b)=aa(ii,jj)*bb(kk,ll)
   return
   end
   subroutine fill(a,b,c,d,aa,aaa,d11,d12,d21,d22)
   integer ii,jj,a,b,c,d,d11,d12,d21,d22,e,f
   real aa(d11,d12),aaa(d21,d22)
   do 44 ii=1,c
   do 44 jj=1,d
   e=a-1+ii
   f=b-1+jj
44 aaa(e,f)=aa(ii,jj)
   return
   end
   subroutine random(seed,randx)
   real randx
   integer seed
   seed=997.0*seed-int(997.0*seed/1.E6)*1.E6
   randx=seed/1.E6
   return
   end
```

The above are subroutines that fill matrices with all elements 0 (foel), fill appropriate elements of a larger matrix with the elements of a smaller matrix (fill), add vectors (add), multiply vectors and tensors (mp), and generate random numbers between 0 and 1 (random).

Once again, Matlab can handle the required matrix mathematics.

5 Connecting the Behaviors/Properties of Polymer Solutions and Liquids to the Microstructural Dependent Conformational Preferences of Their Polymer Chains

INTRODUCTION

The purpose of this chapter is, where possible, to connect the ***Inside*** microstructures and resultant conformational preferences of individual polymer chains to the ***Outside*** behaviors of their solutions and melts. We previously demonstrated in Chapter 4 how the conformational characteristics of polymer chains in the form of their rotational isomeric state (RIS) models can be used to determine the populations of their backbone conformations. The dependence of polymer conformations on their microstructures, together with the dependence of ^{13}C-Nuclear Magnetic Resonance (^{13}C-NMR) resonance frequencies on local backbone bond conformations (γ-gauche nuclear shielding effects), permits assignment of the observed ^{13}C-NMR spectra of polymers to their contributing microstructures. In this chapter, we describe how we may use the conformational preferences of polymer chains to understand the behaviors of polymers in their solutions and melts.

A polymer chain in solution or in its melt is free to adopt its complete range of overall chain conformations, which are summarized and can be evaluated from its RIS conformational model, and is said to be "randomly coiling" (Volkenstein 1963; Birshstein and Ptitsyn 1964; Flory 1969; Tonelli 1986). The spatial volume influenced by a single such randomly coiling polymer, V_i, can be approximated by the volume of a sphere whose radius corresponds to the square-root of its mean-square radius of gyration, R_g, or $V_i = 4\pi R_g^3/3$ (Tonelli 2001, and see Figure 5.1). As pointed out in Chapter 3, $<R_g^2>_o$ and the mean-square end-to-end distances, $<r^2>_o$ of polymer chains can be calculated by matrix multiplication techniques that are appropriate for 1-dimensional systems with pair-wise dependent energies like flexible linear polymers (Ising 1925; Kramers and Wannier 1941; Flory 1969). For sufficiently long and flexible polymers, it is observed that $<R_g^2>_o = <r^2>_o/6$ (Flory 1969).

By comparison to the volume influenced by a randomly coiling polymer, V_i, the volume it actually physically occupies, V_o, or its hard-core volume, is much smaller and given by $V_o = M/N_A\rho$, where M, N_A, and ρ are the molecular weight of the polymer, Avogadro's number, and the bulk polymer density. As we shall shortly see, the ratio V_i/V_o, which we call the polymer chain's intimacy quotient or *IQ*, is typically 100 or larger (Tonelli 2001).

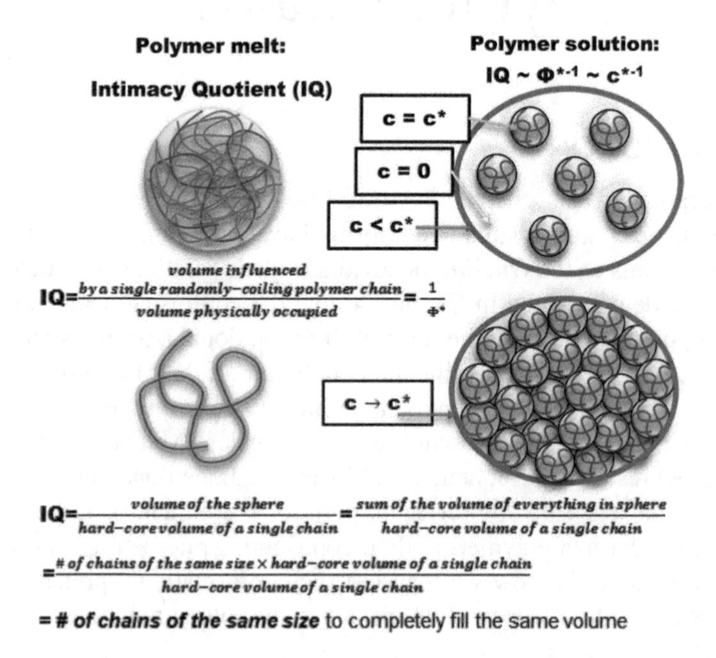

FIGURE 5.1 Definition of IQ and its relationship with the overlapping volume fraction (Φ*) and overlapping concentration (c*). (Reprinted with permission from Shen, J. and Tonelli, A. E., *J. Chem. Educ.*, 94, 1738–1745, 2017. Copyright 2017 American Chemical Society.)

As a consequence, randomly coiling polymer chains are able to influence volumes significantly larger than those they physically occupy, leading to the following conclusions: (1) In solution, a randomly coiling polymer chain can potentially influence a large number or volume of solvent molecules (V_i), much larger than it physically occupies (V_o), suggesting that small amounts (low wt%s) of polymers in solution can lead to large increases in viscosity compared to that of the neat small molecule solvent. (2) In a polymer melt, each randomly coiling polymer chain is in close contact with the repeat units of many other different chains, which can lead to their entanglement.

The characteristic ratio of polymer dimensions, C_n, is defined as $<r^2>_o/nl^2$. The subscript o on $<r^2>_o$ denotes the unperturbed mean-square end-to-end distance, when excluded volume effects are canceled out. nl^2 is the $<r^2>_o$ of a hypothetical freely jointed polymer chain of n bonds each of length l (Flory 1969). In good solvents, polymers have dimensions $<r^2>$ larger than the unperturbed dimensions $<r^2>_o$, because polymer chain coils expand to increase contact with and maximize their favorable interactions with the solvent molecules (Flory 1953). Thus, $<r^2>_o$ can be used as the lower limit for the estimation of $<r^2>$. Though polymer dimensions $<r^2>_o$ generally increase with molecular weight ($\propto$ n and M), C_n achieves a constant asymptotic value C_∞ at sufficiently high M.

As a result, from the value for C_n and M, unperturbed chain dimensions $<r^2>_o$ can be evaluated. $<r^2>_o = C_n\left(nl^2\right) = \left(C_n Ml^2\right)/M_o$, where $M_o = M/n$ is the molecular weight per backbone bond. This yields $V_i = (4\pi/3)\left[\left(C_n Ml^2\right)/6M_o\right]^{3/2}$, which is a lower limit of the pervaded or influenced volume. The number of polymer chains of length n (or mol. wt. M) that would completely fill or occupy the volume pervaded or influenced by a single randomly coiling polymer chain of the same length is the intimacy quotient, or $IQ = V_i/V_o = (4\pi/3)\left[C_n Ml^2/6M_o\right]^{3/2}(\rho N_A/M)$, so $IQ \propto M^{1/2}$.

In Figure 5.1, the entire volume included in the sphere is influenced by a single randomly coiling polymer chain (in red). In a polymer melt, the rest of the space in the sphere is filled with portions of many other polymer chains (in blue) in intimate contact with the red chain. Thus, IQ is an evaluation of the magnitude of the pervaded volume scaled by the hard-core volume of the single chain in red and pictured as the volume included in a tube sheathing the polymer chain. From this figure, IQ is clearly the inverse of the volume fraction of the red chain.

As depicted in the oval on the upper right-hand side of Figure 5.1, in a dilute solution, the red chains are in contact with many solvent molecules (not shown). In its pervaded volume, the volume fraction of a red chain inside the sphere is preserved, while the space outside the spheres is occupied only by solvent molecules. Outside the spheres, the volume fraction of polymer is essentially zero, so the total polymer volume fraction in this dilute solution is smaller than the polymer volume fraction with respect to its pervaded volume (V_o/V_i). The space outside the spheres starts to be filled with other spheres as is depicted in the lower-right oval in Figure 5.1, with an increase in the number of such spheres (increasing polymer concentration). In other words, the polymer volume fractions inside and outside any sphere are equal when the free space outside the spheres is just taken up by other spheres. This is called the onset of the overlap volume fraction, Φ^*, beyond which, polymer chains wander into the pervaded volume of other polymer chains. Φ^* is thus the volume fraction of a polymer chain in a single coil, so IQ = $1/\Phi^*$. The overlap concentration, c*, is a similar parameter and can be calculated with the knowledge of bulk polymer densities. For substantial chain lengths n or molecular weights M, the volume V_i influenced by a flexible randomly coiling polymer chain is much larger (typically ~100 times or more) than the volume V_o physically occupied by its repeat units or segments (hard-core volume).

For example, the [poly(ethylene oxide)](PEO) sample examined in the 1st demonstration of ***Polymer Physics*** presented in Chapter 1 has a M = 4,000,000. The characteristic ratio of PEO is $C_\infty = \langle r^2\rangle_o/nl^2 \sim 4$, n = (4,000,000/44) × 3 = 273,000, $\langle l^2\rangle \sim (2 \times 1.43^2 + 1.54^2)/3 \sim 2.15$ Å², so $\langle r^2\rangle_o = 4 \times 273{,}000 \times 2.15 \sim 2.35 \times 10^6$ Å². Its pervaded or influenced volume $V_i = (4\pi/3)\{[(\langle r^2\rangle_o)^{1/2}]/2\}^3 = 1.9\times10^9$Å³ (Flory 1969). Here, the pervaded volume was considered that of a sphere with a diameter equal to the average end-to-end chain distance $\langle r^2\rangle_o^{1/2}$. Its hard-core occupied volume $V_o = M/\rho N_A = [4 \times 10^6/(1.0 \times 6.023 \times 10^{23})] \times 10^{24}$Å³/cm³ = 6.6 × 10⁶ Å³. This leads to IQ = V_i/V_o = 290. For comparison, the IQ for ethanol is likely not too much greater than 1, and so it is not surprising that 0.5% does not affect the flow time of the 99.5% solvent water molecules.

The PEO polymer coil overlap concentration can be estimated as [$1/(V_i/V_o)$ or 1/IQ] × 100% = 0.34% (to a good approximation, if we take a density 1 g/cm³ for both polymer and water). Thus, a 0.5 wt% solution contains more than enough PEO chains to influence the flow of the entire solution volume, but without significant overlap or entanglement of the PEO chains.

This is why that in all, but the most dilute, solutions, and, of course, in the melt, segments belonging to each polymer chain are in contact with the segments of many other chains, and may become entangled.

INTRINSIC VISCOSITIES OF DILUTE POLYMER SOLUTIONS

We must eliminate polymer chain concentration as a variable to gain information regarding the size, shape, and volume pervaded or influenced by a macromolecule. The specific viscosity, η_{sp}, of a liquid is defined as:

$$\eta_{sp} = \frac{\eta - \eta_o}{\eta_o} = \frac{t - t_o}{t_o},$$

where η and η_o and t and t_o are the viscosities and capillary flow times of non-sheared solutions and solvents (o). The intrinsic viscosity, $[\eta]$, i.e., the contribution to the solution viscosity made by individual, separated, un-overlapped dissolved chains can be obtained by measuring the specific viscosities of a series of polymer solutions and extrapolating to zero concentration.

$$[\eta] = \lim_{c \to 0} \frac{\eta_{sp}}{c}.$$

Einstein suggested that the flow of liquid suspensions of impenetrable beads are characterized by specific viscosities $\eta_{sp} = 2.5(n_2/V)U_e$, where (n_2/V) is the number of solid impenetrable beads or spheres per unit volume and U_e their individual volume (Einstein 1905). Much later Flory suggested for randomly coiling polymer chains, that $U_e = V_i = (4\pi/3)(<r^2>^{1/2}/2)^3$ or $(4\pi/3)(<R_g{}^2>)^{3/2}$ (Flory 1953). This was based on the assumption that as a randomly coiling polymer moves in solution, it retards the motion of the solvent molecules within its sphere of influence. Therefore, the polymer chain and its associated solvent molecules can be thought to move together much like a large impenetrable sphere with a volume V_i, as depicted in Figure 5.2.

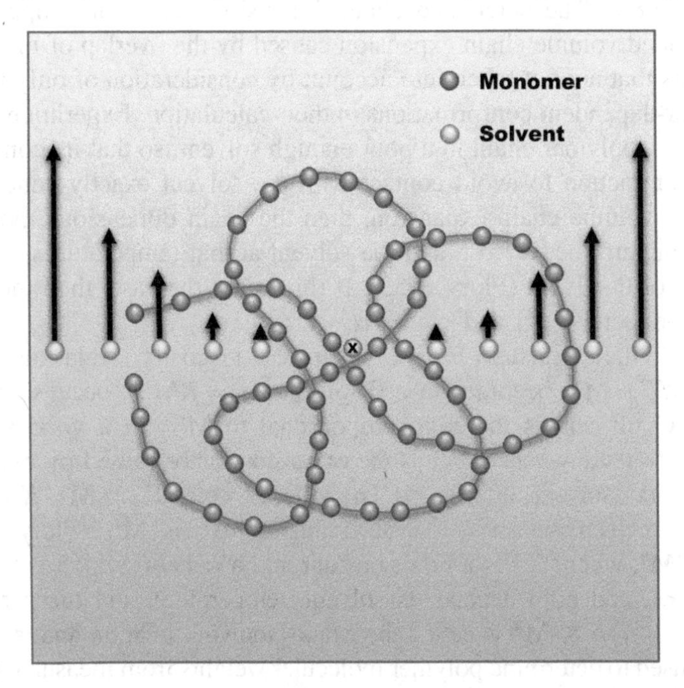

FIGURE 5.2 Translation of a chain molecule with perturbation of the solvent flow relative to the flow of the randomly coiling polymer chain. (Reprinted with permission from Shen, J. and Tonelli, A. E., *J. Chem. Educ.*, 94, 1738–1745, 2017. Copyright 2017 American Chemical Society.)

Since $(n_2/V) = cN_A/100$ M, where c = concentration, in g/100 mL or dL, N_A = Avogadro's number, and M = molecular weight,

$$[\eta] = \lim_{c \to 0} \frac{\eta_{sp}}{c} = \frac{0.025 N_A U_e}{M}.$$

As suggested by Flory, insertion of the explicit expression for $U_e = V_i = (4\pi/3)\left[(\langle r^2\rangle^{1/2}/2)^3\right]$ or $(\langle R_g{}^2\rangle)^{3/2}\Big]$ leads to $[\eta] = \{0.025\, N_A (4\pi/3)\left[(\langle r^2\rangle^{1/2}/2)^3 \text{ or } (\langle R_g{}^2\rangle)\right]^{3/2}\}/M$, which, after combining all constants into a single constant Φ, yields:

$$[\eta] = \Phi\left[\left(\langle \mathbf{r}^2\rangle\right)^{3/2}\right]/\mathbf{M} \propto \boldsymbol{V_i}.$$

$\langle r^2\rangle$ is the mean-square end-to-end distance of a polymer chain dissolved in a particular solvent and at a particular temperature. It is generally not the same as the unperturbed dimensions, $\langle r^2\rangle_o$, calculated with its short-range nearest neighbor-dependent RIS conformational model using the appropriate matrix multiplication techniques (Ising 1925; Kramers and Wannier 1941; Volkenstein 1963; Birshstein and Ptitsyn 1964; Flory 1967, 1969; Tonelli 1986). The calculated chain dimensions $\langle r^2\rangle_o$ are unperturbed by excluded volume chain expansion caused by the overlap of intrachain segments that are not taken into account by consideration of only nearest neighbor-dependent conformations in their calculation. Experimentally, if we place a polymer chain in a poor enough solvent, so that its conformational contraction to avoid contact with the solvent exactly cancels the excluded volume chain expansion, then the chain dimensions measured will be unperturbed $\langle r^2\rangle_o$, and the solvent at that temperatures is called a Flory- or θ-solvent (Flory 1953). If this is not the case, then the chain expansion factor $\alpha > 1$ and $\langle r^2\rangle = \alpha^2 \langle r^2\rangle_o$.

Thus, the Einstein-Flory intrinsic viscosity relation $[\eta] = \Phi[(\langle \mathbf{r}^2\rangle)^{3/2}]/\mathbf{M}$, becomes in a θ-solvent $[\eta] = \boldsymbol{K}\mathbf{M}^{1/2}$, because $\langle r^2\rangle = \langle r^2\rangle_o = C_\infty n l^2$ and is therefore proportional to $M^{1/2}$. In a good solvent, $\langle r^2\rangle = \alpha^2 \langle r^2\rangle_o$, where $\alpha > 1$ is the expansion factor caused by excluded-volume and solvent effects, so $[\eta] = \Phi\left[\left(\alpha^2 \langle \mathbf{r}^2\rangle_o\right)^{3/2}\right]/\mathbf{M} = \boldsymbol{K}\alpha^3\mathbf{M}^{1/2}$. Because α increases with M according to $\alpha^3 = M^{a'}$ (Flory 1953), $[\eta] = \boldsymbol{K}^*\mathbf{M}^{\mathbf{a}}$, where K^* is a new constant and a = ½+a′ which is >½, but usually <1, and both depend on solvent, temperature, and the particular polymer. $[\eta] = \boldsymbol{K}^*\mathbf{M}^{\mathbf{a}}$ is called the Mark-Houwink relation and is conveniently used to determine polymer molecular weights from measured intrinsic viscosities using tabulated values of ***K**** and **a** obtained from the intrinsic viscosities measured in a particular solvent for chemically identical polymers with known molecular weights obtained independently, for example, by light-scattering (Houwink 1954; McGoury and Mark 1954).

Because $[\eta] = \Phi\left[\left(\alpha^2 < r^2 >_0\right)^{3/2}\right]/\mathbf{M} = K^*\mathbf{M}^a$, we may not only determine a polymer's molecular weight M from intrinsic viscosity measurements, but also the quality of the solvent for the polymer $\boldsymbol{\alpha}^2$. If the polymer's RIS conformational model has been developed, then we may calculate its unperturbed dimensions $<\mathbf{r}^2>_0$ that correspond to its measured molecular weight. With knowledge of its molecular weight and dimensions, we can solve $[\eta] = \Phi\left[\left(\alpha^2 < r^2 >_0\right)^{3/2}\right]/\mathbf{M}$ for $\boldsymbol{\alpha}^2$, which is characteristic of the solvent quality for the polymer.

POLYMER ENTANGLEMENT

Let's examine Figure 5.3, where the zero-shear melt viscosities (η) of nine different polymers are plotted *versus* X_w, a parameter proportional to their molecular weight. While below a certain molecular weight M_c, the slope of

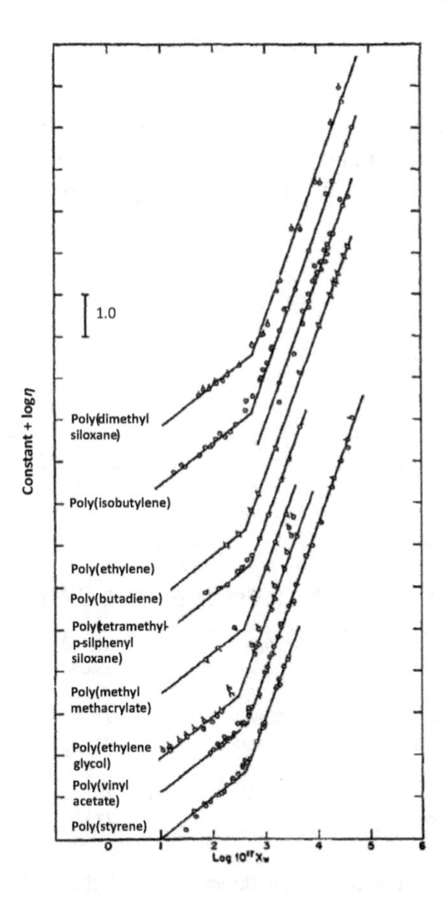

FIGURE 5.3 Log-log plot of the zero-shear melt viscosity as a function of chain-length or molecular weight (X_w α M α n). (Reprinted with permission from Berry, G. C. and Fox, T. G., *Adv. Polym. Sci.*, 5, 261, 1968.)

each plot is proportional to their M, they each change abruptly to a slope of 3.4 above M_c that is different for each polymer. This behavior is believed to be the result of entanglements between polymer chains (see Figure 5.4), which are formed when they are sufficiently long. Similar behavior is shown schematically in Figure 5.5, where, in addition to zero-shear melt viscosities, the molecular weight dependencies of stress relaxation moduli and creep compliances are also presented. All three of these dynamic polymer melt

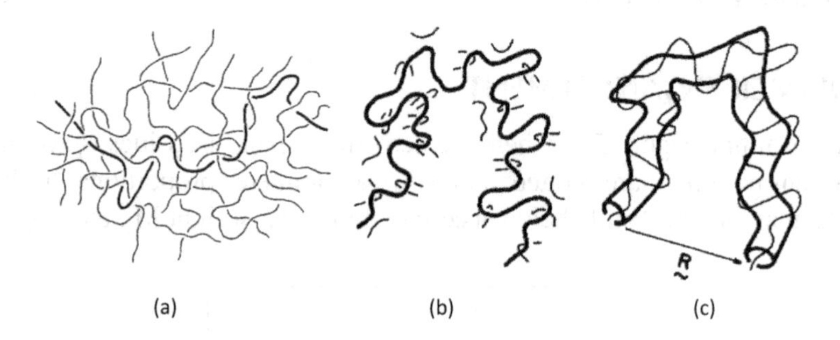

FIGURE 5.4 An entangled polymer melt (a), a chain and segments (b), and a tube of uncrossable constraints both due to entanglement with other chains (c). (Adapted from Graessley, W. W., *Adv. Polym. Sci.*, 47, 67, 1982.)

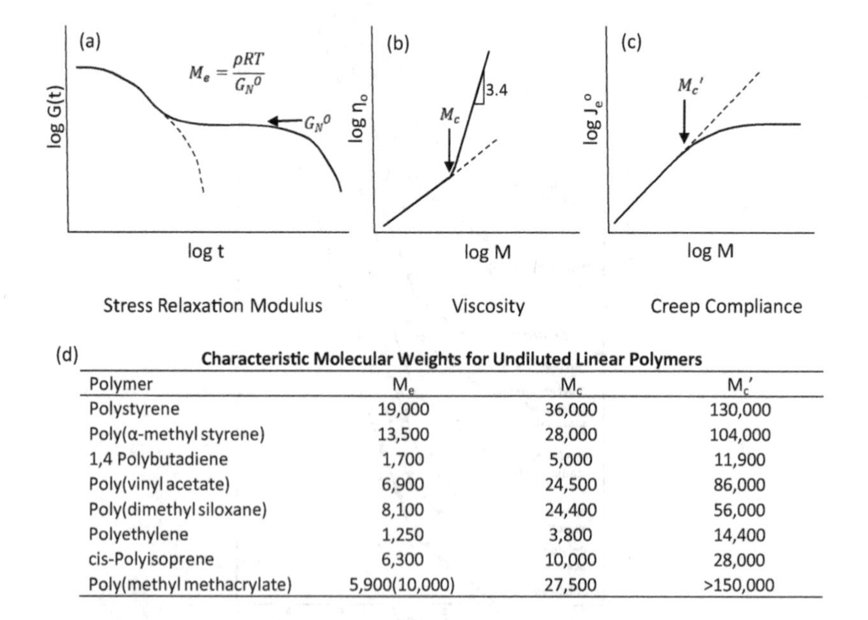

Characteristic Molecular Weights for Undiluted Linear Polymers

Polymer	M_e	M_c	M_c'
Polystyrene	19,000	36,000	130,000
Poly(α-methyl styrene)	13,500	28,000	104,000
1,4 Polybutadiene	1,700	5,000	11,900
Poly(vinyl acetate)	6,900	24,500	86,000
Poly(dimethyl siloxane)	8,100	24,400	56,000
Polyethylene	1,250	3,800	14,400
cis-Polyisoprene	6,300	10,000	28,000
Poly(methyl methacrylate)	5,900(10,000)	27,500	>150,000

FIGURE 5.5 The molecular weight dependence of three characteristic dynamic properties of polymer melts. The stress-relaxation modulus (a), the zero-shear melt viscosity (b), and the creep compliance (c). Molecular weights where each property changes abruptly is tabulated in (d). (Graessley, W. W., *Adv. Polym. Sci.*, 47, 67, 1974.)

behaviors show an abrupt change at a relatively well defined polymer chain length or M. The respective molecular weights where this occurs, however, vary with the property measured [compare the collected data below the plots in Figure 5.5 and note that M_e, M_c, and M_c' differ (Graessley 1974)].

Nevertheless, they are all believed to reflect the effect of entanglements between polymer chains. Entanglements are topological constraints placed on a polymer chain in the melt by the other entangled chains, because they cannot pass through each other (see Figure 5.4). Consequently, they restrict motions that are perpendicular to the contours of the chains (Edwards 1967). Rather, snake-like reptilian motions along the chain contours were suggested for entangled polymer melts (de Gennes 1971, 1979, 1983). In fact, someone actually conducted a study of the analogous reptilian motion of garter snakes (Gans 1970). The garter snakes were faced with moving through a maze of fixed obstacles. As the density of obstacles was increased, it was observed that the force exerted by the snakes to move forward along their body contour increased dramatically.

A theory was developed to describe the reptilian motion of a polymer chain in its melt (Doi and Edwards 1978), and it showed that the zero-shear melt viscosity of polymers should scale as $\mathbf{M^3}$. Subsequent refinements of this theory predict that $\eta \propto \mathbf{M^{3.5}}$ in close agreement with the results shown in Figures 5.3 and 5.5 for $M > M_c$ (Marruci and deCindo 1980; Marruci and Hermans 1980). Because in its melt, the friction factor for a polymer chain with $M < M_c$ should be a simple sum of the friction factors for each of its segments or repeat units, its zero-shear melt viscosity should scale with and be linearly proportional to its M.

Now, we would like to examine the potential connection between a polymer's microstructure and its entanglement molecular weight (M_e, M_c, M_c'). Remember that V_o = volume physically occupied by a polymer = M/ρNA, while V_i = the volume pervaded or influenced by a randomly coiling polymer = $(4\pi/3)(<R_g{}^2>_o)^{3/2}$, where R_g is the radius of gyration. $C_n = <r^2>_o/nl^2$ and $<r^2>_o = 6<R_g{}^2>_o = C_n(nl^2) = C_nMl^2/M_b$, and $M_b = M/n$. Thus, $V_i = (4\pi/3)[(C_nMl^2)/6M_b]^{3/2}$.

The number of polymer chains of length n (or mol. wt. M) that would occupy and completely fill the volume pervaded or influenced by a single randomly coiling polymer chain of the same length is IQ, or the intimacy quotient, $= V_i/V_o$, or $IQ = (4\pi/3)[(C_nMl^2)/6M_b]^{3/2}(\rho N_A/M)$, so $\underline{IQ \propto M^{1/2}}$. Notice that the unperturbed dimensions $<r^2>_o$ and $<R_g{}^2>_o$ were used in deriving the expression for IQ. This is because in bulk amorphous polymers there is no excluded volume expansion of chain dimensions due to the equivalence between intrachain and interchain interactions (Flory 1953). Thus, as mentioned previously, for any substantial chain length n or molecular weight M, the volume V_i influenced by a flexible randomly coiling polymer chain is much larger (by several orders of magnitude) than the volume V_o physically occupied by its repeat units or segments (hard-core volume). This means that in all but the most dilute solutions and, of course, in the molten bulk, segments belonging to each polymer chain are in contact with the segments of many other chains, or they are entangled.

If we assume that all polymers entangle at the same IQ, then we may begin to understand why different polymers begin to become entangled at different chain lengths or Ms, as evidenced by the onset of a plateau in the modulus or creep compliance or a transition of their zero-shear melt viscosities from $\eta \propto M$ to $\eta \propto M^{3.4}$, when $M \geq M_e$ or M_c' or M_c (see Figure 5.5). Assuming IQ(A) = IQ(B) for polymers *A* and *B* at the onset of their entanglements, i.e., at M_e, then the expected ratio of their entanglement molecular weights is $M_e(A)/M_e(B) = [C_n(B)M_b(A)/C_n(A)M_b(B)]^3 \times [\rho(B)/\rho(A)]^2$. As an example, let's compare M_es of PE and polystyrene (PS) in this manner. PE —> $C_n = 6.8$, $M_b = 14$, and $\rho = 0.80$ g/cm^3, and PS —> $C_n = 10.0$, $M_b = 52$, and $\rho = 1.01$ g/cm^3. As a result, $M_e(PS)/M_e(PE) = [(6.8 \times 52)/(10.0 \times 14)]^3 \times [(0.855)/(1.05)]^2 = \underline{\mathbf{12}}$. Experimentally, from the melt viscosity and plateau modulus data in Figure 5.5, $M_e(PS)/M_e(PE) = \underline{\mathbf{14–15}}$. (For an alternative treatment, see Fetters et al. 1994.)

Thus, from the microstructurally sensitive conformational characteristics of individual, isolated polymer chains (RIS models), which lead directly (*via* matrix multiplication methods) to estimates of their sizes, dimensions ($<r^2>_o$, $<R_g^2>_o$, C_n), we are able to understand the onset of chain entanglement in their molten bulk samples. This is ***Polymer Chemistry***!!!

Further emphasizing the importance of the role played by ***Polymer Chemistry***, i.e., the microstructures of polymers, we mention here a paper by Rottach et al. They modeled the diffusion of simple small molecule penetrants in polymer melts by performing a molecular dynamics simulation on a collection of freely jointed chains ($<r^2>_o = nl^2$) to which a few small molecules were added, and whose movements were tracked (Rottach et al. 1999). They found results qualitatively similar to those for diffusion through small molecule liquids. However, when they replaced the freely jointed chains with freely rotating chains by constraining the backbone valence angles to adopt a nearly fixed value, the experimentally observed diffusion behavior of small molecules in polymer melts was reproduced.

More recently Meyer et al. compared molecular dynamics simulations of bead-spring chain melts to that of single randomly-walking chains ($<r^2> = l^2C_\infty n = a^2N^*$, where n and l are the number and length of backbone bonds, and C_∞ is the characteristic ratio of the polymer chain. a is the statistical segment or Kuhn length of the corresponding random walk of $N^* = N/C_\infty$ steps with the same contour length $lN = aN^*$]). Their purpose was to compare self-entanglements or knots formed by single randomly-walking model chains with those found in the more realistic bead-spring melts. They found that the number of knots/self-entanglements in the bead-spring chain melts was in agreement with experimental observations, but nearly an order of magnitude smaller than those expected for the randomly-walking model chain.

This serves to illustrate the potential danger of losing all connections to the distinguishing features of ***Polymer Chemistry*** through adoption of artificial conformational models for polymer chains. Their study also demonstrated that the adoption of over simplified polymer chain models may as well adversely affect the ***Physics*** of polymers needed to understand the unique behaviors of their materials.

An example in this regard is provided by a 50th anniversary perspective appearing in the journal *Macromolecules* that was titled "Polymer Conformation – A Pedagogical Review" (Wang 2017). Professor Wang got off to a good start in his abstract, when he said, "The study of the conformation properties of macromolecules is at the heart of polymer science. Essentially all physical properties of polymers are manifestations of the underlying polymer conformations or otherwise significantly impacted by the conformation properties. In this Perspective, we review some of the key concepts that we have learned over nearly eight decades of the subject and outline some open questions." Unfortunately, the subject he was referring to was ***Polymer Physics***, not polymer conformations, as advertised in his title and the first portion of his abstract. He made this clear as he continued… "The topics include both familiar subjects in ***Polymer Physics*** textbooks and more recent results or not-so-familiar subjects, such as non-Gaussian chain behavior in polymer melts and topological effects in ring polymers. The emphasis is on understanding the key concepts, with both physical reasoning and mathematical analysis, and on the interconnection between the different results and concepts." However, the "key concepts" he was referring to are those of ***Polymer Physics*** not polymer conformations.

As we have repeatedly pointed out here, it is ***Polymer Chemistry*** or the detailed microstructures of polymers that directly affect their conformational preferences, and these, as we have demonstrated, may in many instances be connected to their behaviors and material properties. ***Polymer Physics***, as is evident in Wang's review, depends only marginally upon the microstructurally sensitive conformations of polymers, because he assumes that conformationally polymer chains are either freely jointed with bond lengths $\boldsymbol{b}_0$ or are freely rotating worm-like chains with persistent links of length $\boldsymbol{b} = \boldsymbol{C}_\infty{}^{1/2}\boldsymbol{b}_o$. In other words, in his discussion of ***Polymer Physics***, the conformational characteristics of all polymers are reduced to those approximated by freely jointed or freely rotating model chains, regardless of their detailed microstructures. This, despite the well known sensitivity of their behaviors to their microstructurally dependent conformations.

The conformational characteristics of real polymer chains can rarely if ever be described by freely jointed or freely rotating chain statistics. Rather, their adoption essentially provides only unrealistic approximations of their true conformational ensembles. However, as we have demonstrated, the development of RIS conformational models for polymers, coupled with matrix multiplication techniques applicable to 1-dimensional systems with nearest neighbor-dependent energies, permits a realistic treatment of their conformational characteristics. Importantly, this enables rigorous treatment of many conformationally sensitive properties and behaviors of chemically distinct polymers and leads to the establishment of relevant structure-property relations for their materials.

DYNAMIC BEHAVIORS OF POLYMER SOLUTIONS AND MELTS

We close this chapter by acknowledging that understanding the dynamic behaviors and properties of polymer solutions and melts is generally not possible following the ***Inside*** polymer chain microstructure ←→ ***Outside*** polymer material property approach taken here. The principal reason is that the interactions between polymer chains and between polymer chains and solvents dominate polymer chain dynamics, resulting in an essentially insoluble many-body problem. Thus, for example, polymer rheology in all its guises cannot presently be treated in terms of the ***Inside*** conformational preferences of individual polymer chains. In fact, even adopting simplified artificial models of polymer chains does not really lead to significant understanding of their dynamic behaviors. One need only look at the use of complicated spring and dash pot models of polymer chains that are used to attempt to mimic/reproduce the rheological responses of polymer solutions and melts to appreciate their complexity and confirm our present inability to understand them (Mezger 2014).

REFERENCES

Berry, G. C., Fox T. G. (1968), *Adv. Polym. Sci.*, 5, 261.
Birshstein, T. M., Ptitsyn, O. B. (1964), *Conformations of Macromolecules*, translated by Timasheff, S. N. and Timasheff, N. J. from the 1964 Russian Ed., Wiley-Interscience, New York.
de Gennes, P. (1971), *J. Chem. Phys.*, 55, 572.
de Gennes, P. (1979), *Scaling Concept in Polymer Physics*, Cornell University Press, Ithaca, NY.
de Gennes, P. (1983), *Phys. Today*, 36, 33.
Doi, M., Edwards, S. F. (1978), *J. Chem. Soc. Faraday Trans. II*, 74, 1789, 1802, 1818.
Edwards, S. F. (1967), *Proc. Phys. Soc.*, 92, 9.
Einstein, A. (1905), *Ann. Physik.*, 17, 549.

Fetters, L. J., Lohse, D. J., Richter, D., Witten, T. A., Zirkel, A. (1994), *Macromolecules*, 27, 4639.
Flory, P. J. (1953), *Principles of Polymer Chemistry*, Cornell University Press, Ithaca, NY.
Flory, P. J. (1967), *J. Am. Chem. Soc.*, 89, 1798.
Flory, P. J. (1969), *Statistical Mechanics of Chain Molecules*, Wiley-Interscience, New York.
Gans, C. (1970), *Sci. Am.*, 222, 82.
Graessley, W. W. (1974), *Adv. Polym. Sci.*, 16, 1.
Graessley, W. W. (1982), *Adv. Polym. Sci.*, 47, 67.
Houwink, R. (1954), *Journal für Praktische Chemie (Leipzig)*, 157, 15.
Ising, E. (1925), *Z. Phys.*, 31, 253.
Kramers, H. A., Wannier, G. H. (1941), *Phys. Rev.*, 60, 252.
Marruci, G., deCindo, G. (1980), *Rheol. Acta*, 19, 68.
Marruci, G., Hermans, J. J. (1980), *Macromolecules*, 13, 380.
McGoury, T. E., Mark, H. (1954), *Phys. Meth. Org. Chem.*, 1, 2399.
Meyer, H., Horwath, E., Virnau, P. (2018), *ACS Macro Lett.*, 7, 757.
Mezger, T. G. (2014), *The Rheology Handbook*, 4th ed., Vincentz Network, Hanover, Germany.
Rottach, D. R., Tillman, P. A., McCoy, J. D., Plimpton, S. J., Curro, J. G. (1999), *J. Chem. Phys.*, 111, 9822.
Shen, J., Tonelli, A. E. (2017), *J. Chem. Educ.*, 94, 1738–1745.
Tonelli, A. E. (1986), *Encyclopedia of Polymer Science and Engineering*, 2nd ed., Wiley, New York, Vol. 4, p. 120.
Tonelli, A. E. (2001), *Polymers from the Inside Out*, Wiley-VCH, New York.
Volkenstein, M. V. (1963), *Configurational Statistics of Polymer Chains*, translated by Timasheff, S. N. and Timasheff, N. J. from the 1964 Russian Ed., Wiley-Interscience, New York.
Wang, Z.-G. (2017), *Macromolecules*, 50, 9073.

DISCUSSION QUESTIONS

1. Why are the viscosities of polymer solutions and melts so sensitive to their molecular weights?
2. Describe what happens to the radius of gyration and mean-square end-to-end distance of a polymer chain dissolved in a good solvent as its concentration is increased and give reasons why?
3. We suggested that the molecular weights where chain entanglements begin or the molecular weights between entanglements may be understood on a relative basis through consideration of a polymer's $IQ = V_i/V_o$. Describe the details of this suggestion.
4. Why are we not able to understand the time-dependent dynamic behaviors of polymer solutions and melts nor offer a prognosis for an eventual solution to this difficulty?

6 Connecting the Behaviors/Properties of Polymer Solids to the Microstructural Dependent Conformational Preferences of Their Individual Polymer Chains

INTRODUCTION

As they are in their bulk liquids, polymer chains in bulk solid samples are in close contact, but they may be clearly differentiated by their arrangements and packing and/or mobilities and dynamic behaviors. In polymer solids, both conformational sampling and irreversible movements of their constituent chains are prevented. Amorphous bulk solid samples contain randomly coiled and interpenetrating polymer chains, as do their bulk liquids. However, they differ in the inability of their chains to change conformations and to move irreversibly past each other, i.e., by their chain dynamics not their structural organizations. Semi-crystalline polymer solids, on the other hand, differ drastically from their corresponding bulk liquids and amorphous solids in both their structural organizations and the dynamics of their chains. In crystalline sample regions, polymer chains are generally well organized in single, highly extended rigid conformations that permit them to pack closely together in crystalline registry. At the same time, however, they are in close contact with only a small number of their neighboring crystalline chains (extremely low IQs [intimacy quotients]), albeit over much longer chain portions. As a result, the interactions between crystalline chains are stronger than those between chains in their amorphous regions. The polymer chains in the non-crystalline regions are principally randomly coiled in a fashion similar to those in entirely amorphous samples. In both cases, these amorphous chains may gain mobility when warmed above their glass-transition temperatures (T_g). However, while all the chains in amorphous samples, and those in amorphous regions of semi-crystalline polymer samples, may begin to change their conformations and move above T_g, those in a semi-crystalline sample cannot undergo irreversible flow past each other until the melting temperatures (T_m) of their crystalline regions are reached. Only then can they become completely homogeneous liquid polymer melts, as are wholly amorphous polymers when heated above their glass-transition temperatures.

When stressed, we still expect solid polymer samples to respond principally by altering their conformations, because when strained, they cannot otherwise move past each other. Their necessary conformational responses, however, may be retarded or even precluded by interchain interactions. When under stress, we can expect an internal conformational response, with the speed, or frequency, and the size or amplitude of the response dominated by cooperative interactions between chains and not just the intrachain constraints to internal backbone bond rotations.

As a consequence, behaviors of solid bulk polymer materials that are time or frequency-dependent are not likely to be typically rationalized or explained from the sole perspective of the microstructurally dependent conformational preferences of their individual polymer chains. However, in this case "typically" does not necessarily mean never. Instead, we will describe several solid polymer behaviors, which can at least be rationalized in a comparative manner for chemically different polymers based on their intrachain rotational isomeric state (RIS) conformational models.

SOLID POLYMER PROPERTIES AND Z_{conf}

In Chapter 3, we described how the conformational partition function, Z_{conf}, of a polymer chain may be obtained from its RIS conformational model through the utilization of matrix multiplication of its statistical weight matrices (*Us*) (Flory 1967 1969; Tonelli 1986). Here, we use Z_{conf} to obtain conformational energies and entropies of single polymer chains and attempt to rationalize some properties of solid polymers through their use (Hill 1960; Tonelli 2001).

COPOLYMER T_gs AND THEIR COMONOMER-SEQUENCE DEPENDENCE

We begin our examination of the behaviors of solid polymers with the T_gs of amorphous polymers. While a variety of related structural variables have generally been cited as being important to determining the absolute values of polymer T_gs, the following are often cited: (**1**) the inherent conformational flexibilities of individual polymer chain backbones; (**2**) the sizes, steric bulk, and the relative flexibility of their side-chains; and (**3**) the interactions (steric, dipolar, hydrogen-bonding, van der Waals, etc.) between polymer chains (Shen et al. 2017). However, it is generally agreed, as mentioned later, that polymer T_gs are principally determined by factor (**3**), the interactions between polymer chains. Any structural factor that is altered will affect how polymers interact and, thus, their resultant T_gs. Though (**1**), the inherent conformational flexibility of a polymer chain may influence its T_g, it is usually not the dominating factor. Consequently, we have not as yet attempted to correlate the absolute magnitudes of polymer T_gs, which range over hundreds of °C, as seen in Table 6.1, with their microstructures and the resultant conformational characteristics of their chains.

However, when the T_gs of a series of amorphous aliphatic copolyesters synthesized from propylene and hexamethylene diamines and succinic and adipic acids were recently measured (see Figure 6.1), we found that they were linearly correlated with both the ratio of ester bonds to CH_2 groups (structural factor **3**) and $1/S_{conf}$ (structural factor **1**) (Shen et al. 2017). It remains to be seen if the T_gs of other polymers can also be correlated with their intrachain conformational entropies. Instead, we will show that the comonomer-sequence dependence of copolymer T_gs can be related to their conformational entropies, S_{conf}, which of course can be estimated from $\mathbf{Z_{conf}}$, i.e., $S_{conf} = \mathbf{R\ln Z + (RT / Z)(dZ / dT)}$ (Hill 1960).

The glass-transition temperatures, of homogeneous (A/B) copolymers sometimes depend exclusively on their chemical composition (comonomer weight fractions, W_A, W_B), as described in $1/T_{gp} = (W_A / T_{gA}) + (W_B / T_{gB})$, usually called the Fox equation (Fox and Flory 1950; Fox 1956; DiMarzio and Gibbs 1959), where T_{gA} and T_{gB} are the T_gs of the homopolymers poly-A and poly-B. However, some copolymer T_gs may additionally depend on the sequence distribution of their comonomers, and are given by: $1/T_{gp} = \left[W_A P_{AA} / T_{gAA}\right] + \left[(W_A P_{AB} + W_B P_{BA}) / T_{gAB}\right] + \left[W_B P_{BB} / T_{gBB}\right]$,

TABLE 6.1
Selected Polymer Glass-Transition Temperatures

Polymer	Repeating Unit	Glass-Transition Temperature (°C)
Silicone rubber	$-Si(CH_3)_2-O-$	−125
Polybutadiene	$-CH_2-CH{=}CH-CH_2-$	−85
Polyisobutylene (butyl rubber)	$-C(CH_3)_2-CH_2-$	−70
Natural rubber	$-CH_2-C(CH_3){=}CH-CH_2-$	−70
Polychloroprene (neoprene rubber)	$-CH_2-C(Cl){=}CH-CH_2-$	−50
Poly(vinyl chloride) (P.V.C.)	$-CH_2-CH(Cl)-$	80
Poly(methyl methacrylate) rubber	$-CH_2-C(CH_3)(C{=}O-O-CH_3)-$	100
Polystyrene	$-CH_2-CH(C_6H_5)-$	100

Source: Tonelli, A.E., *Polymers from the Inside Out*, Wiley-VCH, New York, 2001.

where P_{AA}, $P_{AB} = P_{BA}$, P_{BB}, and T_{gAA}, T_{gAB}, and T_{gBB} are, respectively, the populations of and glass-transitions temperatures contributed by each distinct comonomer diad (Johnston 1976).

As observed and shown in Figure 6.2, methyl acrylate (MA) copolymers with vinyl chloride (VC) or vinylidene chloride (VDC), illustrate both behaviors (Hirooka and Kato 1974). VC/MA copolymers follow the linear relation with comonomer weight fraction (Fox and Flory 1950; Fox 1956). Even though the T_gs of PMA and PVDC homopolymers are nearly

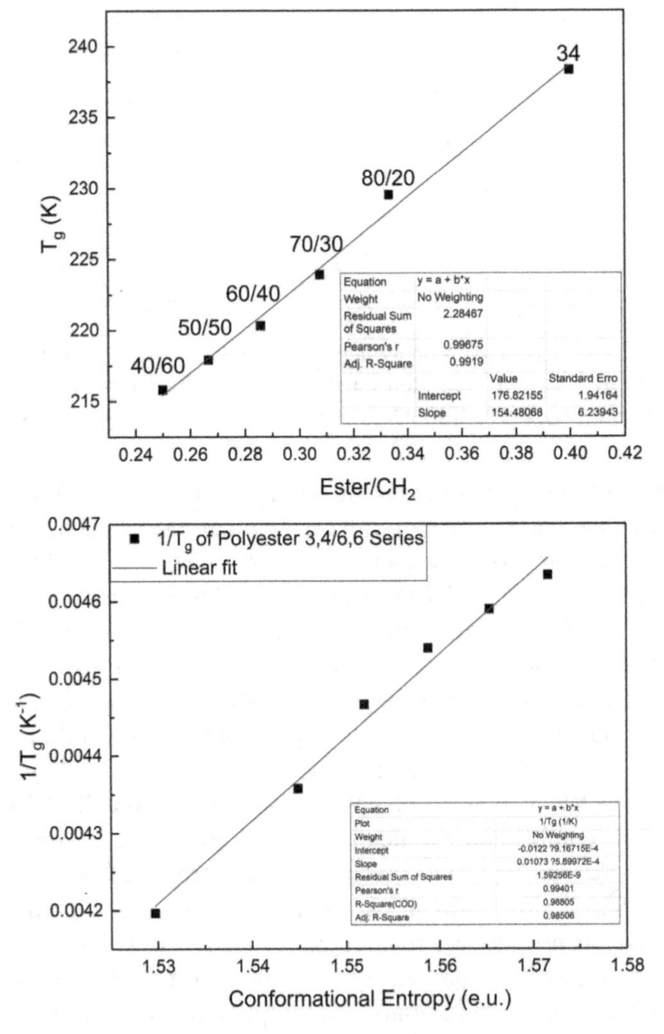

FIGURE 6.1 T_gs of amorphous aliphatic 3,4/6,6 polyesters vs. ratio of ester bonds: CH_2 groups (top) and S_{conf} (bottom). (Adapted with permission from Shen, J. et al., *Polymer*, 124, 235.)

identical, VDC/MA copolymer T_gs deviate strongly from a linear dependence on comonomer weight fractions, with observed T_{gp}s far above those given by the Fox equation.

MA: $-CH_2-CH(-C(=O)-O-CH_3)-$

VC: $-CH_2-CH(Cl)-$

VDC: $-CH_2-C(Cl)_2-$

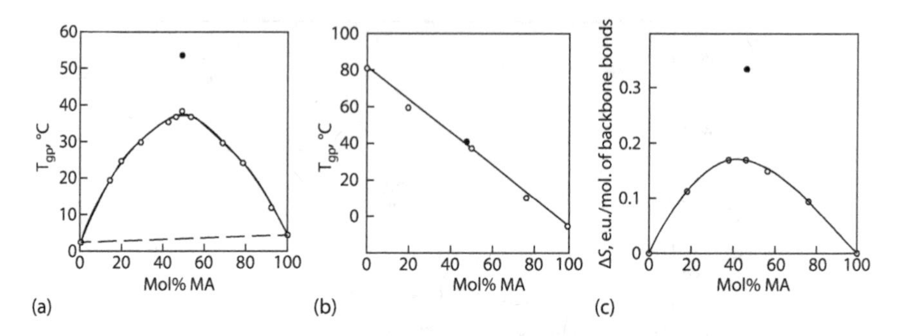

FIGURE 6.2 Experimental copolymer T_gs of VDC/MA (a) VC/MA (b) and calculated trend for VDC/MA (c). (Reprinted with permission from Tonelli, A.E. et al., *Macromolecules*, 43, 6912, 2010. Copyright 2010 American Chemical Society.)

Comparison of the T_{gp}s of random and regularly alternating VDC/MA copolymers, each having the same overall 50:50 molar composition, makes their dependence on comonomer sequence even more obvious. The 50:50 random and regularly alternating VDC/MA copolymers have, respectively, T_{gp}s < 40 and ~55°C, which are significantly elevated from their closely similar homopolymer glass-transition temperatures (~5°C), and that are strictly the result of different contributions made by their distinct comonomer sequences.

We previously postulated that the conformational flexibilities of individual copolymer chains were the source of the distinct T_{gp} behaviors evidenced by VC/MA and VDC/MA copolymers (Tonelli 1974a, 1975, 1977a). The conformational flexibilities of copolymers were assumed to be characterized by their conformational entropies, S_{conf}, as obtained from their RIS conformational partition functions, Z_{conf} according to the usual statistical mechanical relation $S_{conf} = R\ln Z + (RT/Z)(dZ/dT)$, (Hill 1960). The following relationship was then proposed to describe the deviation between the glass-transition temperatures observed for copolymers T_{gp}(obs) and those predicted by the Fox relation, T_{gp}(Fox), which are dependent solely upon the weight-average comonomer composition: $\mathbf{T_{gp}(obs) - T_{gp}(Fox)} \propto \Delta S_{\mathbf{conf}}$, where $\Delta S_{conf} = [(\mathbf{X_A S_A + X_B S_B}) - \mathbf{S_{A/B}}]$ and X_A and X_B are the mole fractions of A and B comonomer units and S_A, S_B, and $S_{A/B}$ are, respectively, the conformational entropies of the homopolymers poly-A and poly-B and their A/B copolymers.

As seen experimentally above in Figure 6.2b and calculated below in Table 6.2, respectively, $T_{gp}(obs) - T_{gp}(Fox)$ and the ΔS_{conf}s calculated for VC/MA copolymers are essentially independent of both comonomer composition and sequence distribution, i.e., they are ≈ 0 ($S_{VC\text{-}MA}$, S_{VC}, and S_{MA} of atactic 50:50 VC-MA, PVC, and PMA are closely similar at 0.741, 0.740, and 0.725 e.u./mole, respectively). Thus, it is expected that $T_{gp}(obs) \approx T_{gp}(Fox)$.

ΔS_{conf}s calculated for VDC/MA copolymers are substantially > 0, on the other hand, with magnitudes dependent on both comonomer compositions and sequence distributions, as seen in Table 6.2 and Figure 6.2c. The conformational entropies calculated for atactic 50:50 VDC/MA, Poly-VDC, and atactic Poly-MA are 0.579, 0.754, and 0.732 e.u./mole, respectively (see Table 6.2). Clearly $\Delta S_{conf} = (X_A S_A + X_B S_B) - S_{A/B} > 0$

TABLE 6.2

Conformational Entropies Calculated for VDC/MA and VC/MA Copolymers

Copolymer	Mole % of MA	Stereoregularity[a]	Comonomer Seq. Distribution	Conformational Entropy S (e.u./mol of backbone bonds)
VDC-MA	0			0.754[b]
VDC-MA	20	A	Random	0.636[b,d]
VDC-MA	40	A	Random	0.579[b,d]
VDC-MA	50	A	Random	0.579[b,d]
VDC-MA	50	A	Alternating	0.412[b,d]
VDC-MA	50		Alternating	0.404[b]
VDC-MA	50	I	Alternating	0.420[b]
VDC-MA	60	A	Random	0.591[b,d]
VDC-MA	80	A	Random	0.641[b,d]
VDC-MA	100	S		0.569[b]
VDC-MA	100	A		0.732[b]
VDC-MA	100	I		1.012[b]
VC-MA	0	S		0.845[c]
VC-MA	0	I		0.900[c]
VC-MA	0	A		0.740[c,d]
VC-MA	50	S	Alternating	0.768[c]
VC-MA	50	I	Alternating	0.854[c]
VC-MA	50	A	Random	0.741[c,d]
VC-MA	100	S		0.560[c]
VC-MA	100	I		1.015[c]
VC-MA	100	A		0.725[c,d]

Source: Tonelli, A.E., *Macromolecules*, 8, 54, 1975.

[a] A atactic, I isotactic, and S syndiotactic.

[b] Calculated at 5°C.

[c] Calculated at 40°C.

[d] Average of ten Monte Carlo generated chains, where the mean deviation from the average entropy is ≈1%–2% (Tonelli 1975).

for VDC/MA copolymers (Tonelli 1975). Furthermore, the conformational entropies calculated for atactic 50:50 random and regularly alternating VDC/MA copolymers (0.579 vs. 0.412 e.u./mole in Table 6.2) make it apparent ΔS_{conf} and T_{gp} for regularly alternating VDC-MA are both expected to be greater than those of the 50:50 random VDC/MA copolymer.

The comonomer and stereosequence dependence of styrene-acrylonitrile (S/AN) copolymers and styrene-methylmethacrylate copolymers were also treated successfully in terms of $T_{gp}(\text{obs}) - T_{gp}(\text{Fox}) \propto \Delta S_{conf}$ (Tonelli 1974a, 1977a). Consider the experimental observations of the glass-transition behavior of syndiotactic S/AN copolymers, in which the phenyl ring and C≡N side-chains have alternating positions along the copolymer's fully extended backbone. They have shown T_gs that are lower than those predicted by the Fox equation (Tonelli 1974a, 1975). It has also been observed that the glass-transition temperatures of isotactic S/AN copolymers, in which the side-chains are all located on the same side of the fully extended polymer backbone, are greater than those predicted by the Fox relation. Finally, experimental observations have led to the conclusion that the glass-transition temperature of regularly alternating isotactic S-AN copolymers is higher than that of a 50:50 random isotactic S-AN copolymer with the same comonomer composition.

Conformational entropies were calculated from RIS conformational modeling using matrix multiplication methods for S/AN copolymers with varying comonomer-sequence distributions and stereoregularities, and are presented in Table 6.3.

TABLE 6.3

Conformational Entropies of S/AN Copolymers Calculated Using the RIS Matrix Multiplication Method

Sequence Distribution	Mole Fraction of Styrene Units	Stereo-regularity	S, e.u./mol of Backbone Bonds
Regularly alternating	0	Isotactic	1.323
Regularly alternating	50	Isotactic	1.137
Random	50	Isotactic	1.148
Regularly alternating	100	Isotactic	1.015
Regularly alternating	0	Syndiotactic	1.381
Regularly alternating	50	Syndiotactic	1.225
Random	50	Syndiotactic	1.198
Regularly alternating	100	Syndiotactic	1.006

Source: Tonelli, A.E., *Macromolecules*, 7, 632, 1974a.

From these calculated conformational entropies several conclusions can be reached (Tonelli 1974a, 1977a). First, syndiotactic S/AN copolymers have higher calculated conformational entropies and so are presumably more flexible than the weighted average of syndiotactic Poly-S and Poly-AN homopolymers. Since they are more flexible, we would expect the syndiotactic copolymer to have a lower T_g than would be predicted by bulk additive relations, such as the Fox equation. Second, it is clear that isotactic S/AN copolymers have lower conformational entropies and are presumably less flexible than the weighted average of isotactic poly-S and poly-AN homopolymers. Based on this observation, it is expected that the isotactic copolymer would have a higher T_g than that predicted by bulk additive relations, such as the Fox equation.

Finally, for isotactic 50:50 S/AN copolymers, the calculated conformational entropy of the random copolymer is greater than for the regularly alternating copolymer. Thus, we expect that random 50:50 isotactic S/AN copolymers will be more flexible and have a lower T_g than the regularly alternating isotactic copolymer. On the other hand, the conformational entropy calculated for a random 50:50 syndiotactic S/AN copolymer is less than that of a regularly alternating syndiotactic S/AN copolymer, suggesting that the regularly alternating syndiotactic S/AN copolymer is more flexible and should have a lower T_g.

Indeed, all of these Tg predictions drawn from the calculated conformational entropies of S/AN copolymers mirror those that are experimentally observed (Tonelli 1974a, 1975, 1977a).

As a final example of copolymer T_gs, consider those of atactic styrene-p-Br-styrene (S/p-BrS) copolymers (Tonelli et al. 2010). As mentioned previously in Appendix 4.1, atactic S/p-BrS copolymers were produced by brominating atactic poly-S in solvents of different quality (α). The S/p-BrS copolymers obtained in solvents of good, θ, and poor quality for poly-S were expected to yield increasingly blocky samples, because of the increasing difficulty in brominating their interior repeat units as their random-coil conformations became more compact (Semler et al. 2007). The glass-transition temperatures observed for these S/p-BrS copolymers and their poly-S and poly-p-BrS homopolymers are presented in Figure 6.3. There, a linear dependence on the composition of comonomers typical of Fox equation behavior is clearly indicated, with no dependence on comonomer sequence.

The observed adherence of S/p-BrS copolymer T_gs to Fox-like behavior is expected. We mentioned in Appendix 4.1 that poly-S and poly-p-BrS homopolymers and their S/p-BrS copolymers, all with the same tacticity, should possess closely similar conformational characteristics. This is because the p-bromine substituents on the phenyl rings are too far removed from the polymer backbone to affect their backbone conformations (Yoon

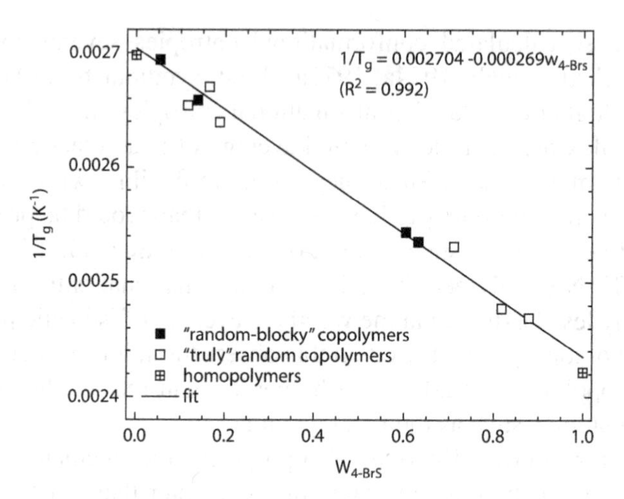

FIGURE 6.3 T_gs of atactic S/*p*-BrS copolymers having "truly" random (open symbols) and "random-blocky" (closed symbols) comonomer-sequence distributions. (Reprinted with permission from Tonelli, A.E. et al., *Macromolecules*, 43, 6912, 2010. Copyright 2010 American Chemical Society.)

et al. 1975). In other words, their T_gs should only depend on their comonomer compositions and tacticities, but not upon their comonomer sequences.

Williams and Flory (Williams and Flory, 1967; Flory, 1969) have shown that the conformations of neighboring methylene sequences in aliphatic polyesters, like those in the 3,4/6,6 aliphatic polyesters mentioned previously are conformationally independent. This is a consequence of their separation by the two ester bonds. Consequently, their conformational entropies should be independent of comonomer sequences and depend only on comonomer composition, suggesting that their T_gs should be described by the Fox equation in agreement with the results in Figure 6.1.

In summary, it appears that the observed deviations away from linear composition vs T_g, or Fox behavior, observed in copolymers can be understood in terms of whether or not the weighted conformational entropies calculated for their constituent homopolymers are equal to or deviate from the conformational entropies of their resultant copolymers. This is clearly another successful example of the ***Inside*** polymer chain microstructure ←→ ***Outside*** polymer material property approach.

MELTING TEMPERATURES OF SEMI-CRYSTALLINE POLYMERS

Just as critical as T_gs are to the processing and use temperatures of amorphous polymers, so are the melting temperatures, T_ms, of semi-crystalline polymers. As seen in Table 6.4, the T_ms of the crystalline regions in semi-crystalline polymers depend sensitively on their microstructures and range

TABLE 6.4

Polymer Melting Temperatures, Entropies, and Enthalpies

Polymer	T_m, °C	ΔS_m[a]	ΔS_{exp}	$(\Delta S_m)v = \Delta S_m - \Delta S_{exp}$	$\Delta S_{conf.}$	ΔH_m	$1/\Delta S_m$
Polyethylene	140(413 K)	2.29–2.34	0.46–0.52	1.77–1.84	1.76	960	0.43
Isotactic polypropylene	208(481 K)	1.50	0.44–0.65	0.85–1.09	0.96	720	0.67
cis-1,4-polyisoprene	28(301 K)	0.87	0.45	0.43	1.00	260	1.15
trans-1,4-polyisoprene	74(346 K)	2.19	0.91	1.28	1.37	760	0.46
Polyoxymethylene	183(456 K)	1.75	0.35	1.40	1.50	820	0.57
Polyoxyethylene	66(339 K)	1.78	0.37	1.41	1.70	600	0.56
Polyethyleneterephthalate[b]	267(540 K)	1.46	0.29	1.17	1.07	790	0.68
Polytetrafluoroethylene	327(600 K)	1.97	0.52	1.45	1.60	1180	0.51
cis-1,4-polybutadiene	5(278 K)	1.92	0.43	1.49	1.38	530	0.52
Polyethyleneadipate	65(338K)	1.48	0.38	1.10	1.04	500	0.68
Polyethylenesuberate	75(348 K)	1.50	0.38	1.12	1.16	520	0.67
Polyethylenesebacate	83(356 K)	1.54	0.38	1.16	1.24	550	0.65
	<450 K>					<720>	<0.58>

[a] All entropies are given in e.u./mol of backbone bonds.

[b] Phenyl ring is treated as a single bond.

over hundreds of °C. Here, we explore whether the conformational characteristics of polymers, which are also sensitively dependent on their microstructures, may be related to the T_ms of semi-crystalline polymers.

Melting is a 1st-order thermodynamic phase transition between ordered solid crystalline and disordered liquid phases, with both coexisting in equilibrium only at the melting temperature. It is clearly a structural transition, because the arrangements and packing, as well as the mobilities, of molecules in crystalline solids and amorphous liquids are so disparate. This is in contrast to the glass-transitions of amorphous polymers, which involve a change in molecular mobility, but not in molecular structure.

In Figure 6.4, the temperature-dependent specific volume of amorphous atactic polystyrene (a-PS) is compared with that of semi-crystalline polyethylene (PE), as the former passes through its T_g and the latter its T_m. Below and above T_g, the structure of a-PS remains unchanged, while the a-PS chains begin to sample different conformations and move irreversibly relative to each other above T_g. This simply results in an increased sensitivity to temperature of the a-PS specific volume (Figure 6.4a). Melting of a semi-crystalline polymer like PE, on the other hand, results in an abrupt discontinuous change in specific volume at T_m, as well as distinct temperature dependencies below and above.

Figure 6.5 describes schematically our thermodynamic analysis of polymer melting (Tonelli 1970, 1974b). We hypothesize an additional thermodynamic state intermediate to the crystalline (c) and molten amorphous (a) states and call it the "expanded" crystalline state. In the proposed "expanded" crystalline state, polymer chains remain in their crystalline conformations, but are separated and packed more loosely in an expanded volume equal to the volume of the melt (V_a). The transition between the crystalline and "expanded" crystalline states is strictly an interchain transition from a total crystalline volume V_c to V_a, the volume of the melt in the expanded crystal, accompanied by no change in the crystalline chain conformations. The internal conformational energies and entropies of the crystalline polymer chains (E_c and S_c) remain the same following the volume expansion to the "expanded" crystalline state. Because of the volume expansion, however, the interchain energy and entropy are altered, because this is an exclusively interchain transition.

The only changes produced in the final constant volume transition between the "expanded" crystalline and melt states are the accompanying changes in polymer chain conformations, their conformational energies ($E_a - E_c$), and entropies ($S_a - S_c$), i.e., this is solely an intrachain transition. We can readily evaluate E_a and S_a from the RIS conformational partition function Z_{conf}, i.e., $E_a = (RT^2/Z)(dZ/dT)$ and $S_a = R\ln Z + (RT/Z)(dZ/dT)$ (Hill 1960). Because polymers usually crystallize in their single lowest energy conformation (Flory 1969), E_c and S_c are both typically 0. Thus,

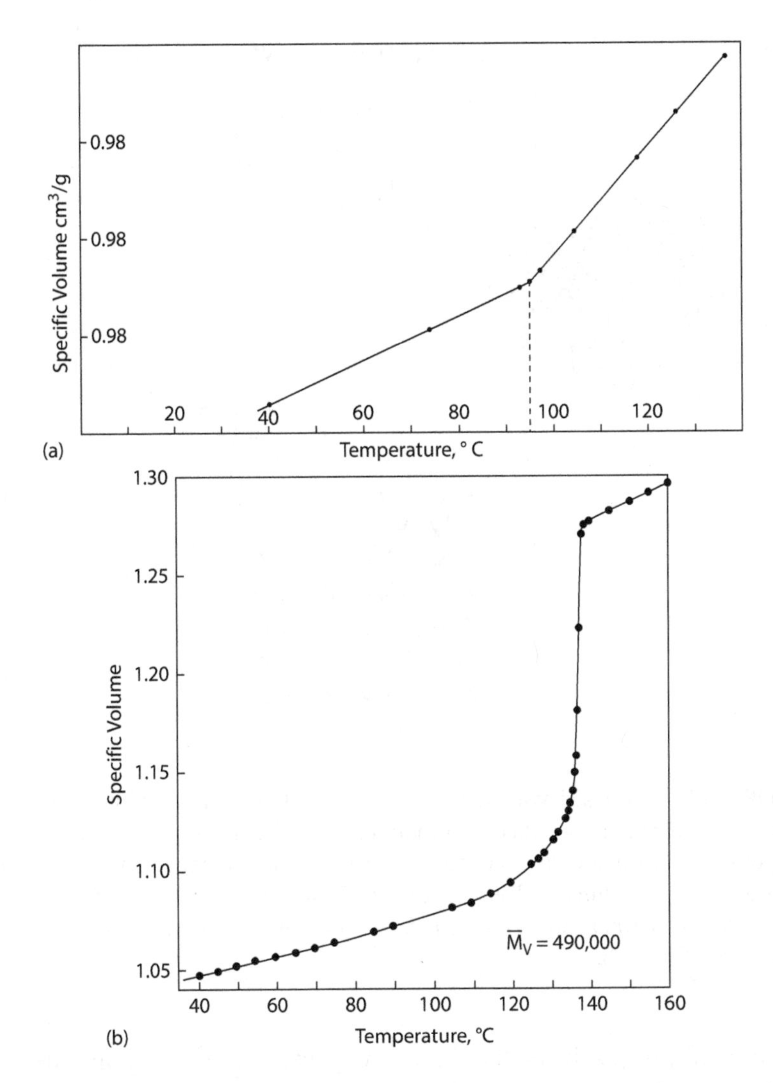

FIGURE 6.4 The change in specific volume with temperature of (a) amorphous atactic polystyrene (Adapted with permission from Gordon, M. and Macnab, I.A., *Trans. Faraday Soc.*, 49, 31–39, 1953.) and (b) semi-crystalline polyethylene. (Reprinted with permission from Chiang, R. and Flory, P.J., *J. Am. Chem. Soc.*, 83, 2857, 1961. Copyright 1961 American Chemical Society.)

changes in energy and entropy for the transition between the "expanded" crystalline and molten states are reduced to the conformational energy E_a and entropy S_a of single randomly coiling molten polymer chains.

The change in entropy accompanying the transition from the crystal to the "expanded" crystal, ΔS_V or ΔS_{exp}, is caused by the expansion of the sample volume (ΔV_m), so ΔS_V or $\Delta S_{exp} = \Delta V_m (\partial S / \partial V)_T \cdot (\partial S / \partial V)_T = (\partial P / \partial T)_V$ (a Maxwell relation, Niven 1890) and $(\partial P / \partial T)_V = (\alpha / \beta)$, where

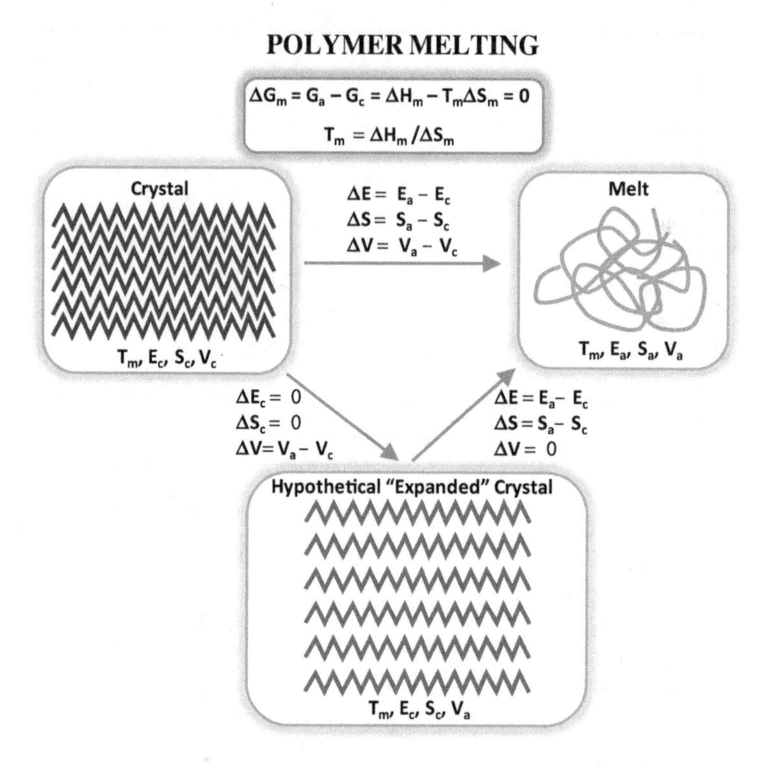

FIGURE 6.5 Proposed two-stage melting of polymer crystals:interchain volume expansion and intrachain conformational disordering. Note: $E_{a,c}$ and $S_{a,c}$ refer only to the conformational contributions made by individual chains and $V_{c,a}$ are the whole sample volumes. (Tonelli, A.E., *J. Chem. Phys.*, 52, 4749, 1970; Tonelli, A.E., *Analytical Chemistry*, vol. 3, R.S. Porter and J.F. Johnson, Eds., Plenum, New York, 89, 1974b.)

α and β, are, respectively, the thermal expansion coefficient and the isothermal compressibility both at T_m (measured on the melt just above T_m). As a result, $\Delta S_{exp} = \Delta V_m(\alpha/\beta)$ (see Appendix 6.1 for a detailed derivation).

The change in entropy accompanying the transition from the "expanded" crystal to the melt, which is a constant volume process, is $(\Delta S_m)_v = \Delta S_m - \Delta S_{exp}$ = constant volume entropy of fusion, which should, as we pointed out above, be the same as ΔS_{conf}, i.e., $(\Delta S_m)_v = \Delta S_{conf}$. Comparison of the fifth and sixth columns in Table 6.4, above, shows that for 11 of the 12 polymers listed there, in fact $(\Delta S_m)_v$ and ΔS_{conf} are closely similar. The single outlier is natural rubber (cis-1,4-polyisoprene), and the source for the disagreement between its measured $(\Delta S_m)_v$ and its calculated ΔS_{conf} has been investigated and discussed, but is not yet fully resolved (Immirzi et al. 2005).

Notice in Table 6.4 that the total entropy of fusion S_m is dominated by $(\Delta S_m)_v$, which is essentially ΔS_{conf}. By way of a typical example, let us examine the melting of PE to determine the relative magnitudes of the enthalpy and entropy changes accompanying the two transitions in the thermodynamic model of polymer melting shown in Figure 6.5. For PE, T_m = 140°C, ΔH_m = 960 cal/mole, and ΔS_m = 2.3 e.u./mole. The change in conformational energy accompanying melting is $\Delta E_{conf} = E_a - E_c, = E_a$, and the energy of the crystalline all *trans* conformation E_c is 0. In the melt, 40% of the backbone bonds are $g^{\pm}$, so $E_a = 0.4 \times 500$ cal/mole = 200 cal/mole. Ignoring the ΔV_m term, the total melting enthalpy, ΔH_m = 960 cal/mol, 960–200 cal/mole = 760 cal/mole must come from the volume expansion and separation of chains produced on melting. For PE, $\Delta E_{exp} >> \Delta E_{conf}$ and $S_{conf} >> \Delta S_{exp}$. (see columns 3 and 4 in Table 6.4) So as an approximation, we may replace $T_m = \Delta H_m / \Delta S_m$ by $T_m \approx \Delta E_{exp} / \Delta S_{conf}$ = (760 cal/mole)/(1.8 e.u./mole) = 420K or ~150°C, which is indeed close to the experimental T_m = 140°C (413K) of PE (see Figure 6.4).

It appears that the heat of melting polymer crystals ΔH_m is dominated by interchain interactions ($\Delta H_m \sim \Delta E_{exp}$), while the melting entropy ΔS_m is dominated by intrachain conformational flexibility ($\Delta S_m \sim \Delta S_{conf}$). This leads to a **Golden Rule:**

> ***** T_m should be high, low for polymers with strong, weak interchain crystalline forces, which are largely absent in the melt, and with limited, extensive conformational freedom in the melt*****

It should be mentioned, as seen in Table 6.4, that the heats of melting ΔH_m, rather than values of $1/\Delta S_m$, are generally much more closely correlated with the melting temperatures ($T_m = \Delta H_m / \Delta S_m$) of semi-crystalline polymers. However, an important exception to this correlation is discussed below, and serves to close our discussion of polymer melting with an instructive example of our **Inside** polymer chain microstructure ←→ **Outside** polymer material property approach.

Nylons-6, -6,6, and -6,10 and other aliphatic polyamides melt at temperatures much higher (~200°C) than the structurally analogous aliphatic polyesters. However, the heats of fusion ΔH_m for these two classes of crystalline polymers are similar (Tonelli, 1971). Since ($T_m = \Delta H_m / \Delta S_m$), the higher melting temperatures of the nylons must be caused by their smaller entropies of fusion ΔS_m, not their higher heats of melting.

Experimentally determined total entropies of fusion for nylons-6 and -6,6 are 1.2 to 1.3 e.u./mole of backbone bonds, which are significantly reduced from those of the structurally analogous aliphatic polyesters in Table 6.4. Their calculated conformational entropies are 1.7 e.u./mole of backbone bonds, leading to the absurd result that for these aliphatic nylons $\Delta S_{conf} > \Delta S_m$. Absurd because $\Delta S_m = \Delta S_{conf} + \Delta S_{exp}$, and $S_{exp} = \Delta V_m (\alpha/\beta)$

is not negative, because their melts have a lower density and greater volume than their crystals.

It would therefore appear that the $\Delta S_{conf} = S_a - S_c$ calculated for the nylons are too large. We are confident that the RIS conformational models for the aliphatic nylons are appropriate and are closely similar to those of the corresponding aliphatic polyesters, so we believe their calculated S_as are valid (Williams and Flory 1967; Flory 1969). In order to reduce $\Delta S_{conf} = S_a - S_c$, we must assign a non-zero positive value to S_c. Usually polymers crystallize into a single extended conformation, determined to be all *trans* for both the aliphatic nylons and polyesters at room temperature, so that their S_c and E_c can both legitimately be expected to be 0. If, however, in the nylon crystal, the chains develop some conformational disorder before reaching their T_m and melting, then $S_c > 0$ and $\Delta S_{conf} = S_a - S_c$ would be reduced below the observed ΔS_m as it must be.

The crystal structures of nylons-6, -6,6, and -6,10 were examined by Itoh using X-ray diffraction and FTIR spectroscopy over the temperature range 20–240°C (Itoh 1976). Above 100°C, he observed, from the X-ray diffraction of their aligned fibers, a significant crystal contraction along their unit cell c-axis, or chain direction, for all three nylons as shown in Figure 6.6. The nylon repeat unit lengths were shortened by 2.2%–3.9% from the all-*trans* conformation between room temperature and 200°C. By comparison, between 20°C and 120°C, only a 0.18% contraction in the crystalline chain repeat length was observed for PE by Itoh. Crystalline nylon chains are clearly not in their fully extended, all-*trans*, lowest energy conformation when they melt, and S_c therefore ≠, but is > 0.

Using polarized FTIR spectroscopy, Itoh showed that amide group hydrogen bonds between adjacent chains in the crystal are retained up to T_m, and he proposed a conformational model to account for both his X-ray

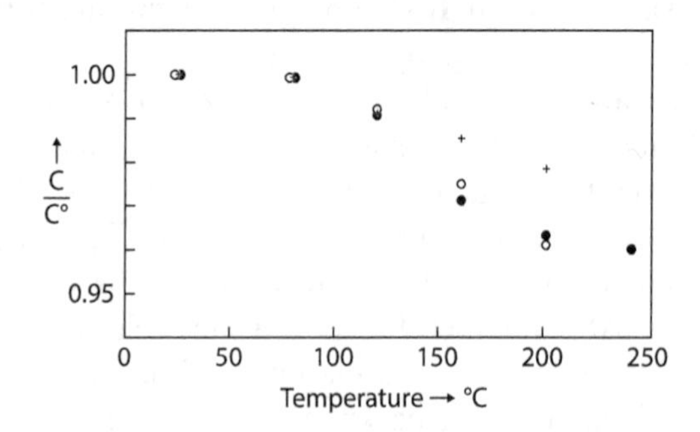

FIGURE 6.6 Temperature dependent c-axis spacings from X-ray diffraction of nylon fibers. The room temperature c-axis spacing is C_o and nylons are –6 (+), –6,10 (o), and –6,6 (•). (Reprinted with permission from Itoh, T., *Japn. J. Appl. Phys.*, 15, 2295, 1976.)

and IR observations. To retain the hydrogen-bonded network, Itoh fixed each amide bond in the planar *trans* conformation, while rotations of $\tau°$ were allowed about all bonds except the -N–CH_2- bond which was rotated $-\tau°$ to maintain the amide groups in the same plane. He determined that rotation angles of 26, 36, and 39° for $|\tau|$ and all -CH_2–CH_2- bond rotations (φs) could reproduce the shortened c-axes observed by X-ray for nylons-6, -6,6 and -6,10, respectively (see Figure 6.7).

However, we showed that this level of conformational disorder led to estimated S_cs of 1.6 and 2.4 e.u./mole for nylons-6 and 6,6, which are much too high (Tonelli 1974b). As a consequence, we modified his model for conformational disordering by fixing all φs in the *trans* conformation. Only τ counter rotations were permitted about the -N(H)–CH_2- and -CH_2–N(H)- and bonds. $|\tau s|$ = 45 and 60° were then required to achieve the chain repeat contractions seen by Itoh in nylons-6 and -6,6, respectively, and corresponded to S_c ~ 0.6 and 0.7 e.u./ mole of backbone bonds. Now $S_{conf} = S_a - S_c$ = 1.7 – (0.6 – 0.7) or 1.1 and 1.0 e.u./mole of backbone bonds, which compares favorably with, i.e.., is less than ΔS_m = 1.2 to 1.3 e.u./mole of backbone bonds.

Thus, the bottom line conclusion is that aliphatic nylons melt at significantly higher temperatures than aliphatic polyesters, not because of hydrogen-bonding between crystalline nylon chains, which are disrupted upon melting, but because of conformational disorder in the crystalline nylon chains occurring at high temperatures, but below T_m. The reason that aliphatic nylons and polyesters melt with similar enthalpies, ΔH_ms, is because a significant number of the hydrogen bonds between neighboring crystalline nylon chains are reformed and replaced in their

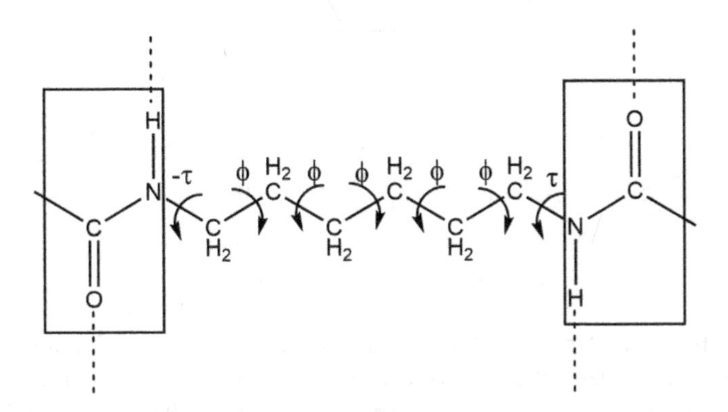

FIGURE 6.7 Itoh's model for the conformational disorder in nylon-6,6 chains. (Itoh, T., *Japn. J. Appl. Phys.*, 15, 2295, 1976.)

randomly coiling melts, albeit not all are formed between different chains (Slichter 1955, 1959). The retention of substantial amide group hydrogen-bonding in nylon melts cause their lower than expected ΔH_ms, which approach those of the structurally analogous polyesters that are not able to form hydrogen-bonds at all.

THE FLEXIBILITIES OF POLYMERS WITH 1,4-ATTACHED PHENYL RINGS IN THEIR BACKBONES

POLY(ETHYLENE PHTHALATES)

We begin with poly(ethylene terephthalate) (PET), whose structure is drawn in Figure 6.8. Williams and Flory developed a RIS conformational model for PET and concluded that the bond conformations on either side of the 1,4-phenyl ring, i.e., ϕ_1 and ϕ_2 rotations, are independent of each other (Williams and Flory 1967). This means that the *trans* and *cis* arrangements of carbonyl groups are not only the preferred conformations, but they are also equally populated. As a consequence, they concluded that the phenyl ring can be considered a freely rotating link, **L**, in the statistical sense, because the net average rotation angle φ is 0 or 180° with equal probability.

Subsequently, we investigated the dynamic behavior of the phenyl rings in PET using the semi-empirical energy functions described below to estimate the energies, $V_{\phi 1,2}$, required to rotate (ϕ_1 and ϕ_2) around the phenyl to carbonyl carbon bonds on either side of each phenyl ring, which, remember, are independent of each other (Tonelli 1973a).

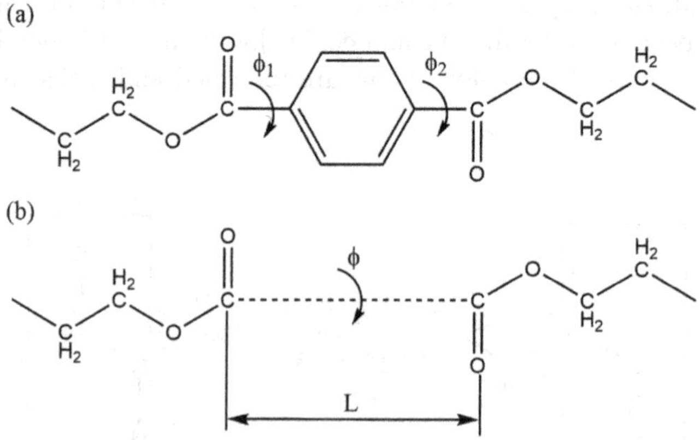

(a)

(b)

FIGURE 6.8 (a) Portion of PET chain encompassing a terephthaloyl residue with a *trans* arrangement of carbonyl groups, where $\phi_1 = \phi_2 = 0°$. (b) PET with the phenyl ring in the terephthaloyl residue replaced by the virtual bond L, and the net rotation angle $\varphi = \phi_1 + \phi_2 = 0°$. (Tonelli, A.E., *J. Polym. Sci., Polym. Lett., Ed.*, 11, 441, 1973a.)

$$V_{\phi_{1,2}} = V_{nb} + V_{\pi}$$

$$V_{nb} = \sum_{i,j} \frac{A_{ij}}{r_{ij}^{12}} - \frac{C_{ij}}{r_{ij}^{6}}$$

$$V_{\pi} = \left[\left(V_{\pi}^{\circ}\right)/2\right]\left(1-\cos 2\phi_{1,2}\right)$$

V_{nb} is the usual non-bonded van der Waals energy, and V_{π} is the energy required to localize the π-electrons to separately occupy the phenyl ring and the ester group as ϕ_1 and ϕ_2 rotate from their coplanar *trans* and *cis* conformations. Based on the rotation barriers observed in benzaldehyde and several of its para-substituted derivatives, we estimated the π-barrier in PET as $V_{\pi}^{\circ} \sim 5$ kcal/mole (Green et al. 1962; Anet and Ahmad 1964; Silver and Wood 1964; Fateley et al. 1965; Tonelli 1973a).

In Figure 6.9, the calculated conformational energies $V_{\phi 1 \text{ or } \phi 2}$ are presented. The sum of the two independent rotations ϕ_1 and ϕ_2 on either side of each phenyl ring yield the net rotation φ about a terephthaloyl residue virtual bond **L**. This makes the probability of a virtual bond rotational state φ, f_{φ}, the product of the independent probabilities, $(SW_{\phi 1})(SW_{\phi 2})$, of the two independent rotations ϕ_1 and ϕ_2. $f_{\varphi} = \frac{(SW_{\phi_1})(SW_{\phi_2})}{\sum_{\phi_1,\phi_2}(SW_{\phi_1})(SW_{\phi_2})}$, where for example: $SW_{\phi_1} = \frac{\sum_{\phi_2} exp\left[-V(\phi_1,\phi_2)/RT\right]}{\sum_{\phi_1,\phi_2} exp\left[-V(\phi_1,\phi_2)/RT\right]}$.

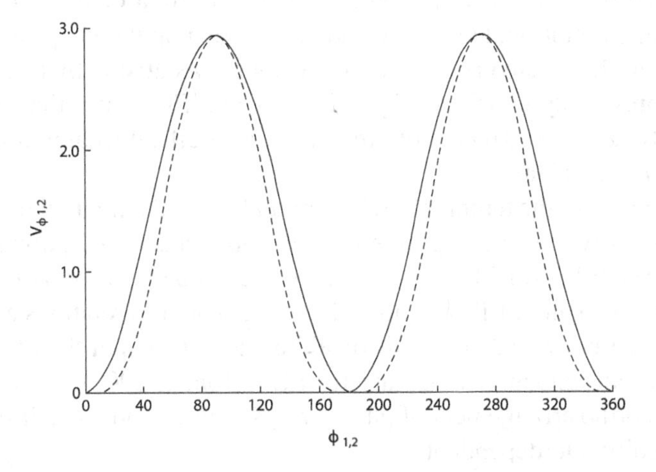

FIGURE 6.9 Estimated energy of rotation $V_{\phi 1 \text{ or } \phi 2}$ about the phenyl to carbonyl carbon bonds in PET (solid line is $V_{nb} + V_{\pi}$ and dotted line is V_{π} with V_{π}° = 2.95 kcal/mole. (Reprinted with permission from Tonelli, A.E., *J. Polym. Sci., Polym. Lett., Ed.*, 11, 441, 1973a.)

TABLE 6.5
Fractional Probabilities, f_φ, of Rotational States about the Virtual Bond L Spanning a Terephthaloyl Residue in PET

Φ(deg)	f_φ
0,180	0.0574
10,170,190,350	0.0550
20,160,200,340	0.0487
30,150,210,330	0.0399
40,140,220,320	0.0302
50,130,230,310	0.0209
60,120,240,300	0.0130
70,110,250,290	0.0076
80,100,260,280	0.0044
90,270	0.0035

Source: Tonelli, A.E., *J. Polym. Sci., Polym. Lett., Ed.*, 11, 441, 1973a.

The fractional probabilities of the virtual bond rotational states f_φ calculated when V_π° = 5.0 kcal/mole and T = 25°C are presented in Table 6.5. Every virtual bond rotational state is appreciably populated, with a maximum energy difference at 25°C of ~1.65 kcal/mole between any two rotational states about the virtual bond **L**. A true dynamically freely rotating bond or link would evidence no rotational barriers, and, therefore, all rotational angles would be equally populated ($f_\varphi = (1/36) = 0.028$). The 1,4-phenyl rings in PET are not only statistically freely rotating, as previously concluded by Williams and Flory, but they are also nearly dynamically freely rotating links as well (Williams and Flory 1967; Tonelli 1973a).

In Figure 6.10 the terephthaloyl, isophthaloyl, and phthaloyl residues in PET, and its two isomers, poly(ethylene isophthalate) and poly(ethylene phthalate) (PEI and PEP), are drawn. We developed conformational RIS models for PEI and PEP and showed that the ϕ_1 and ϕ_2 rotations about the phenyl to carbonyl carbon bonds on either side of each of the PEI phenyl rings are independent, as they are for PET (Tonelli 2002). However, the proximal ortho arrangement of phenyl rings in PEP renders their rotations energetically interdependent.

NMR studies of oriented semi-crystalline fibers and amorphous samples of PET by ^{13}C-NMR, PEI by ^{13}C- and ^{1}H-NMR, and of poly(butylene terephthalate) (PBT) with deuterated phenyl rings by ^{2}H-NMR have been reported (VanderHart et al. 1981; Cholli et al. 1984; Abis et al. 1998).

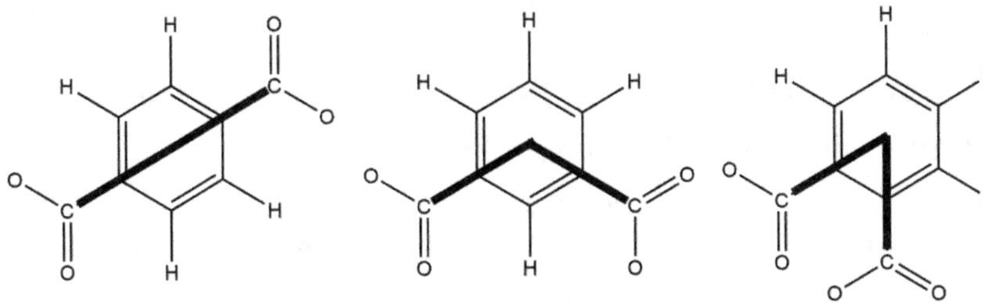

PET PEI PEP

FIGURE 6.10 From left to right the terephthaloyl, isophthaloyl, and phthaloyl residues in PET, PEI, and PEP.

These investigations have demonstrated that the 1,4-phenyl rings in both PET and PBT are dynamically flexible and undergo facile "flipping" between their coplanar *trans* ($\varphi = 0°$) and *cis* ($\varphi = 180°$) conformations. The 1,3-phenyl rings in PEI, on the other hand, were observed to be rigid. This is due to the fact that the geometry of the 1,3-phenyl rings in PEI precludes their "flipping" without moving other portions of their attached chain backbones, as are permitted by their 1,4-attachment in PET and PBT.

Through temperature-dependent ^{2}H-NMR observation of the phenyl rings in the amorphous portions of a deuterated PBT sample, Cholli et al. observed their "flipping" was accompanied by an activation energy of 5.9 kcal/mole (Cholli et al. 1984). This is in close agreement with the 2 × 3.0 kcal/mol = 6.0 kcal/mole phenyl ring "flipping" barrier we expect from Figures 6.8 and 6.9, because both of the approximately 3 kcal/mol barriers that accompany the independent ϕ_1 and ϕ_2 rotations must be simultaneously surmounted during the "flipping" of a 1,4-phenyl ring.

"Flipping" of 1,4-phenyl rings in polymer backbones can have some important macroscopic consequences for their materials. For example, the permeabilities (*P*) of amorphous PET, PEI, and PEP films to oxygen were measured (Polyakova et al. 2001). These authors found that both PEI and PEP films exhibited a dramatic, approximately four-fold, reduction in oxygen permeability in comparison to a PET film. The PEP film exhibited the smallest permeability. Gas permeability (*P*) is the product of the diffusivity (*D*) and solubility (*S*) of the gas through and in the film, i.e., $P = D \times S$. Both quantities should be independently measured before correlations with polymer microstructures are drawn, and Polyakova et al. followed this protocol (Polyakova et al. 2001). They observed oxygen solubilities that were nearly the same in each of the isomeric poly(ethylene phthalate)s and oxygen permeabilities and diffusivities that were proportional. Clearly, oxygen diffuses much more rapidly (~4 times faster) through amorphous PET than either PEI or PEP.

TABLE 6.6
Densities, Crystallinities, CO_2 Diffusion Coefficients, Solubilities, and Permeabilities for Amorphous PET and PEI Films

Polymer	ρ (g/cm³)	X_c (%)	M_∞ (mg CO_2/g polymer)	D × 10^{10}	S × 10^5 (cm³/cm³·Pa)	P × 10^{14} (cm³/cm·s·Pa)
PET	1.344	5	4.20	9.10	4.66	4.24
PEI	1.351	0	1.34	2.24	1.49	0.34

The oxygen permeability and diffusion data reported by Polyakova et al. are consistent with the notion that the diffusion of small penetrant gases in films of the amorphous isomeric poly(ethylene phthalate)s may be dominated by phenyl ring motion, and, in particular, ring "flipping" for PET (Tonelli 2002). In agreement with Abis et al., we suggested that the diffusivity of gases in these poly(ethylene phthalate)s may be controlled by the dynamic flexibilities of their chains rather than the amounts of static free volume in their samples (Abis et al. 1998).

To further test this suggestion, we measured the permeabilities of CO_2 gas through amorphous PET and PEI films (Kotek et al. 2004). The results are displayed in Table 6.6. In agreement with the O_2 diffusivities reported by Polyakova et al., we found CO_2 diffusivity in PET to be four-fold greater than that of the PEI film. On the other hand, we found the solubility of CO_2 in PET was also greater (three-fold greater) than in PEI, while Polyakova et al. found similar O_2 solubilities in their PET and PEI films. The net result of our observations was that the permeability of gaseous CO_2 through PET was ~12 times greater than through PEI. Again, we concluded that this is partially due to the facile "flipping" of the independently rotatable 1,4-phenyl rings in PET that provide effective facile pathways for the diffusion of gases, diffusive pathways unavailable to the 1,3-phenyl rings in PEI.

Polymers with High Impact Strengths Well Below Their Glass-Transition Temperatures

Generally, solid amorphous polymers are brittle well below their glass-transition temperatures. However, some polymers, those with 1,4-linked phenyl rings in their backbones, can exhibit high impact strengths at room temperature, which may be as much as 100–200°C below their T_gs, and even at substantially lower temperatures (Heijboer 1968, 1969; Tonelli 1972, 1973b). An example is poly(2,6-dimethyl-1,4-phenylene oxide) (PPO) shown in Figure 6.11a.

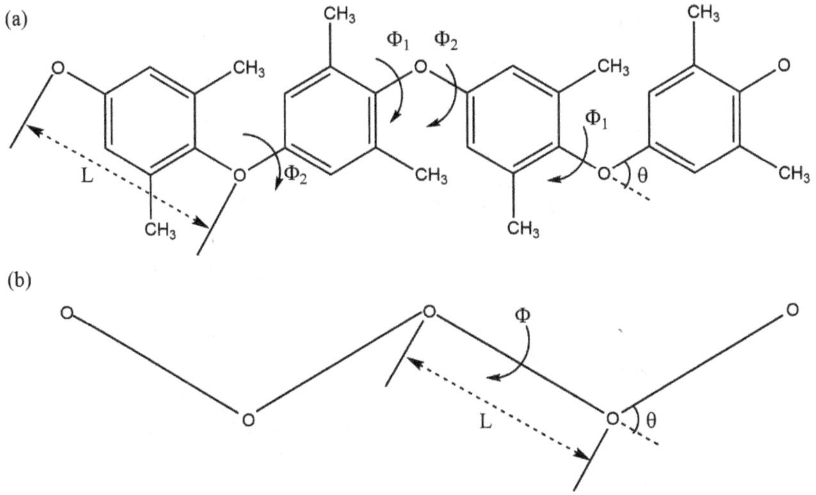

FIGURE 6.11 (a) A portion of PPO chain in the planar zigzag conformation, with all ϕ_1s and ϕ_2s = 0°, and all phenylene rings are coplanar. (b) A portion of the PPO backbone in the planar zigzag conformation, where neighboring ether oxygen atoms are connected by virtual bonds of constant length L. (Tonelli, A.E., *Macromolecules*, 6, 503, 1973b.)

Amorphous polymers that exhibit high impact strengths, especially glassy polymers, must somehow be able to adjust their entangled conformations rapidly enough to avoid being torn apart by the localized high frequency, high strain deformation involved in an impact event. In an attempt to connect high impact strength with the facile conformational dynamics of polymers whose backbone motions experience no or minimal inherent barriers, the conformations of PPOs with H, H; CH_3,CH_3; CH_3, phenyl; and phenyl, phenyl 2,6-substituents were examined (Tonelli 1972, 1973b). All show high impact strengths at room temperatures and below, yet have T_gs of 90, 205, 160–180, and 220°C, respectively.

Our approach was similar to that used on PET (Tonelli 1973a). We replaced the actual PPO chain in Figure 6.11a with a virtual chain of oxygen atoms connected by virtual bonds of constant length **L** seen in (b). While pairs of rotation angles ϕ_2,ϕ_1 on either side of the phenyl rings are independent, ϕ_1,ϕ_2 pairs on either side of each oxygen are not. Their conformational energies $V(\phi_1,\phi_2)$ were estimated in a manner similar to those in PET and are presented in Figure 6.12 (Tonelli 1973b).

The net virtual bond rotation angle φ is the sum of the two independent rotation angles ϕ_2 and ϕ_1 about the real O-C_1 and C_4-O bonds on each side of a phenyl ring. Because their probabilities are independent, the probability f_φ, of a virtual bond angle φ, is just the product of the independent probabilities of rotational states ϕ_2 and ϕ_1, where $\phi_2 + \phi_1 = \varphi$.

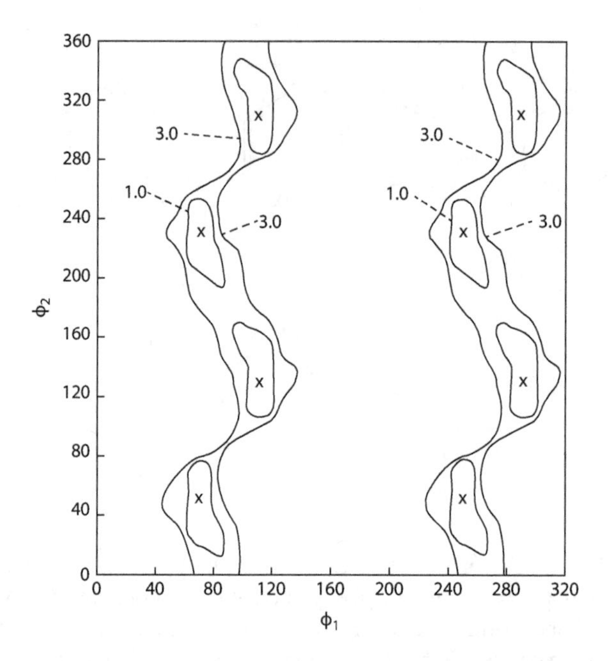

FIGURE 6.12 The 2,6-dimethyl PPO conformational energy, $V(\phi_1,\phi_2)$, map for ϕ_1,ϕ_2 pair conformations on either side of a backbone oxygen atom. Energy contours in kcal/mole of repeat unit are drawn relative to the minimum energy conformations denoted by X's. (Reprinted with permission from Tonelli, A.E., *Macromolecules*, 6, 503, 1973b. Copyright 1973b American Chemical Society.)

$$f_\varphi = (SW_{\phi_2})(SW_{\phi_1}) / \sum_{\phi_1,\phi_2} (SW_{\phi_1})(SW_{\phi_2}),$$

and, for example, the statistical weight or probability of conformations with any particular ϕ_1 is given by:

$$SW_{\phi_1} = \frac{\sum_{\phi_2} exp\left[-V(\phi_1,\phi_2)/RT\right]}{\sum_{\phi_1,\phi_2} exp\left[-V(\phi_1,\phi_2)/RT\right]}$$

Thus, for any ϕ_1 = say Φ, we simply use Figure 6.12 to determine the conformational energies $V(\Phi, \phi_2)$, and then evaluate their corresponding SWs.

Table 6.7 lists the fractional virtual bond rotational probabilities for the most commonly used PPO, 2,6-dimethyl-PPO. Of greatest significance is the uniformity in the distribution of f_φ, which means that every virtual bond net rotational state φ is appreciably accessible. At 25°C, the maximum energy differences between the virtual bond rotational states is 1.4 kcal/mole. For a truly freely rotating virtual bond, $f_\varphi = (1/18) = 0.056$ for all values of the net rotation angle φ. Similar results were obtained for

TABLE 6.7
The Fractional Populations f_φ of the Net Rotational Angle φ Around the Backbone Bonds in 2,6-dimethyl-PPO

φ	f_φ
0	0.036
20	0.063
40	0.060
60	0.073
80	0.036
100	0.036
120	0.073
140	0.060
160	0.063
180	0.036
200	0.063
220	0.060
240	0.073
260	0.036
280	0.036
300	0.073
320	0.060
340	0.063

Source: Tonelli, A.E, *Macromolecules*, 5, 558, 1972.

all the PPO derivatives (Tonelli 1973b), so all the PPO polymers should therefore behave as dynamically freely rotating chains when isolated from other polymer chains, as in their dilute solutions.

Polycarbonate (PC) (see Figure 6.13) and the polysulfones shown below, which have two distinct virtual bonds, i.e., those connecting the backbone oxygens to the quaternary carbons or to the sulfur atoms, should also behave conformationally as dynamically freely rotating chains. Again, this is likely the reason they also exhibit high room temperature impact strengths well below their T_gs (Heijboer 1968, 1969; Tonelli 1972, 1973b).

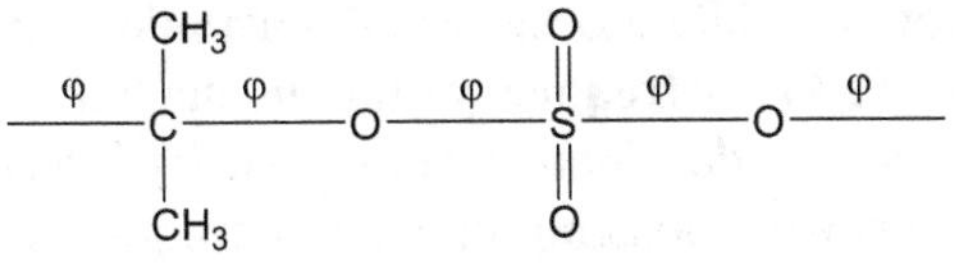

CH3 O
φ φ φ φ φ
—C—O—S—O—
CH3 O

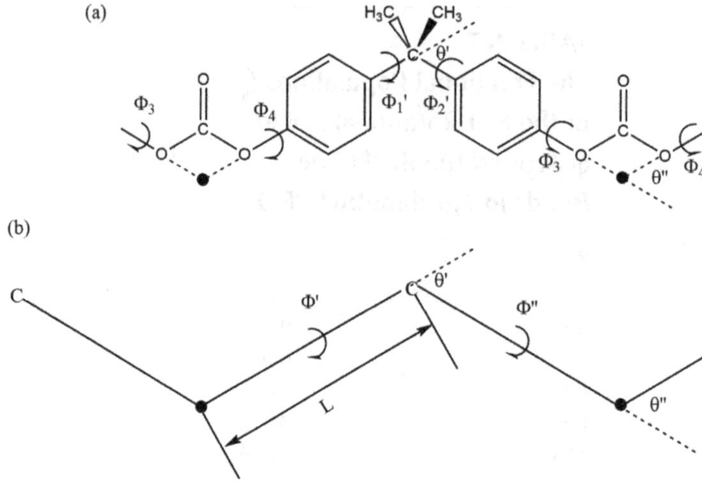

FIGURE 6.13 Polysulfone and (a) PC in the planar zigzag all *trans* conformation, where all phenylene rings are coplanar. (b) Same portion of PC, but with virtual bonds L connecting the quaternary carbons with the •s in (a), and the net virtual bond rotation angles φ' and φ'', which are the respective sums of the independent rotations $\phi_1' + \phi_4$ and $\phi_2' + \phi_3$.

The high impact strength and ductility shown well below the glass transitions for polymers containing 1,4-linked phenyl groups in their backbones may seem puzzling, especially considering their high T_gs. If, however, we assume that the impact strength, a locally high frequency, high amplitude deformation, is related to the ability of a polymer chain to undergo rapid, reversible conformational transitions without having to surmount numerous intrachain barriers that can lead to bond rupture, then the nearly dynamically free rotation nature of polymers containing 1,4-linked phenyl groups in their backbones may be the source of their substantial glassy state impact strengths. Because this class of polymers possess no substantial intramolecular barriers to chain motion, it would appear that their high glass-transition temperatures are primarily the result of intermolecular interactions between their bulky chains. This is confirmed by the distinct T_gs of poly(phenylene oxides) with different 2,6-substituents mentioned above.

These considerations lead to the following **Golden Rule:**

> ****Bulk mechanical properties of polymer materials dependent upon high frequency chain motions, such as impact strength, are primarily governed by intramolecular conformational barriers to polymer motion, while mechanical deformations and other processes involving lower frequency motions, such as polymer glass-transitions, are predominantly controlled by intermolecular or interchain barriers opposing these lower frequency motions.****

We close this discussion by noting that even though PET has an essentially dynamically freely rotating terephthaloyl residue in its backbone, the -O—CH_2-, -CH_2—CH_2-, and -CH_2—O- bonds in the ethylene glycol portions of the PET chain (see Figure 6.8) are only able to adopt *trans* and *gauche*± conformations that are separated by barriers of several kcal/mole (Williams and Flory 1967). They prevent facile intramolecular conformational transitions and likely lead to the relatively brittle room temperature mechanical behavior seen for PET (Shen et al. 2016), even though its 1,4-linked phenyl rings may be able to rapidly flip.

ELASTIC POLYMER NETWORKS

When the long polymer chains in an amorphous sample are cross-linked to form a 3-dimensional network, above its glass-transition temperature, the network will evidence reversible elasticity. The polymer network may be deformed by stretching, for example, but removal of the stretching force is very quickly followed by a reversible recovery of the network's original pre-stretched shape. Everyday examples are provided by rubber bands and rubber balloons that are, respectively, uniaxially stretched or isotropically inflated and then released or allowed to deflate. In both cases, the rubber band and the balloon recover their original sizes and shapes. Such reversible elasticity is a prime example of ***Polymer Physics***, because only networks composed of cross-linked, yet inherently mobile, polymer chains exhibit this behavior.

To form elastic networks that are readily stretched and can reversibly return to their un-stretched states, i.e., to their un-stretched size and shape, when the stretching force is removed, requires three factors: (i) flexible polymer chains, which can alter their sizes and shapes; (ii) polymer chains must be mobile and able to change their conformations when the network is stretched, so the temperature must be higher than the softening temperatures (T_g and T_m) of the amorphous and crystalline regions of the polymer sample; and (iii) the polymer chains must be cross-linked into a 3-dimensional network to eliminate irreversible movements or flow. In reality, a cross-linked polymer network is a single macroscopic molecule (see Figure 6.14). The cross-linking of natural rubber (cis-1,4-polyisoprene) with sulfur at elevated temperatures is called vulcanization and was discovered by Charles Goodyear (1837). An example of vulcanization is depicted schematically in Figure 6.15, where another rubber, poly-1,4-butadiene, is cross-linked with sulfur.

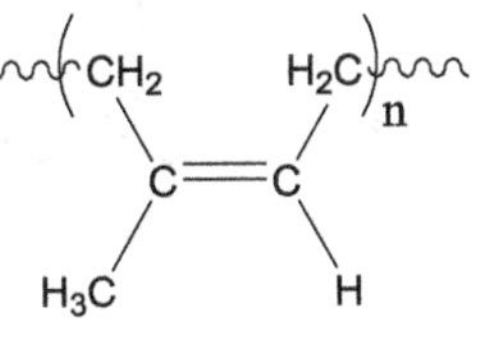

CH_2 H_2C n
C=C
H_3C H

cis-1,4-polyisoprene

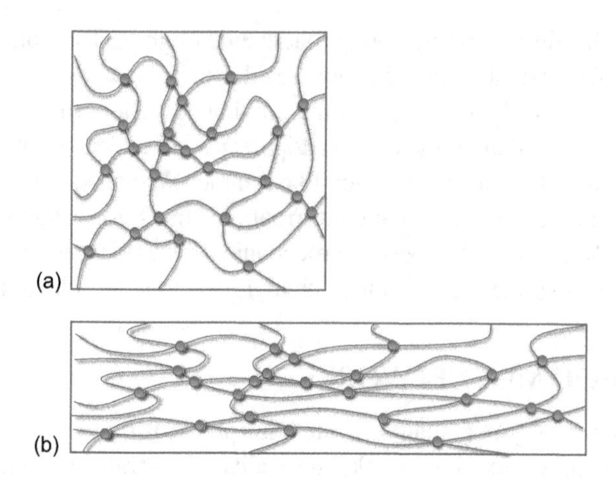

FIGURE 6.14 Cross-linked polymer network in (a) unstrained and (b) strained states. (Reprinted with permission from Shen, J. and Tonelli, A.E., *J. Chem. Educ.*, 94, 1738–1745, 2017. Copyright 2017 American Chemical Society.)

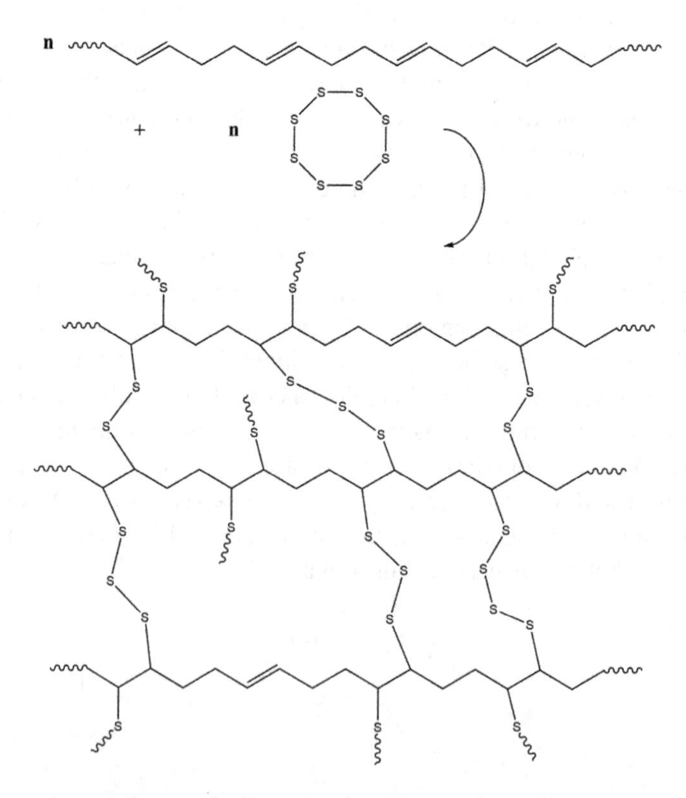

FIGURE 6.15 Vulcanization of poly-1,4-butadiene.

Thermodynamics of Polymer Networks

When a rubber band or any other polymer network is stretched, as indicated in Figure 6.14, some chains between cross-links are extended, while others are compressed. This results in the polymer network maintaining its unstretched volume during stretching, i.e., $\Delta V \sim 0$, so the average distances between polymer chains are maintained and ΔE(interchain) ~ 0, as is ΔS(interchain). In Figure 6.16, we illustrate that the distance between the ends of a short polymer chain may be drastically altered by changing the conformation of a single bond internal to the chain-ends. As a result, we conclude that upon stretching, ΔE(intrachain) ~ 0, because large changes in the distances between cross-links can be achieved by altering a small number (fraction) of backbone bond conformations. Thus, the total ΔE(stretching) ~ 0!

From the first law of thermodynamics: $\Delta E = \delta Q + \delta W \sim 0$, or $\delta Q = -\delta W . \delta W = f\Delta l = f(l - l_o) = -\delta Q$. Because f and $l - l_o$ are both > 0, $\delta Q < 0$, or stretching results in a release of heat, as was mentioned in Chapter 1. From the second law of thermodynamics: $\delta Q/T = \Delta S, \delta Q = T\Delta S$, where $\Delta S = \Delta S_{conf}$, because on stretching, interchain contributions to ΔV, ΔE, and ΔS are all ~ 0. As a result, $\delta Q = T\Delta S_{conf} = -f(l - l_0)$, and $S_{conf} = (-f/T)(l - l_o) < 0$.

Stretching reduces the network entropy, which must be intramolecular, and, because $\Delta V \sim 0$, is therefore conformational. ΔS_{conf} is the source of the network's resistance to being stretched (retractive force f), leading to its reversible retraction upon removing f. $\Delta S_{conf} = (-f/T)(l - l_o)$ or $f = -T\Delta S_{conf}/(l - l_o)$, so <u>f↑ as T↑</u>. The force of retraction f increases with T, and a greater force must be applied to the network in order to maintain its stretched length as the temperature is increased, or the network will contract, $l - l_0 = -T\Delta S_{conf}/f$ (see Figure 6.17).

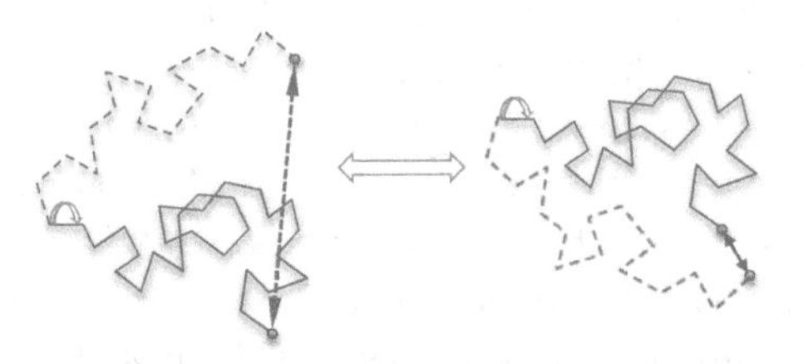

FIGURE 6.16 Illustration of the reduction in the end-to-end separation of polymer chain-ends achieved by altering the conformation about a single backbone bond. (Reprinted with permission from Shen, J. and Tonelli, A.E., *J. Chem. Educ.*, 94, 1738–1745, 2017. Copyright 2017 American Chemical Society.)

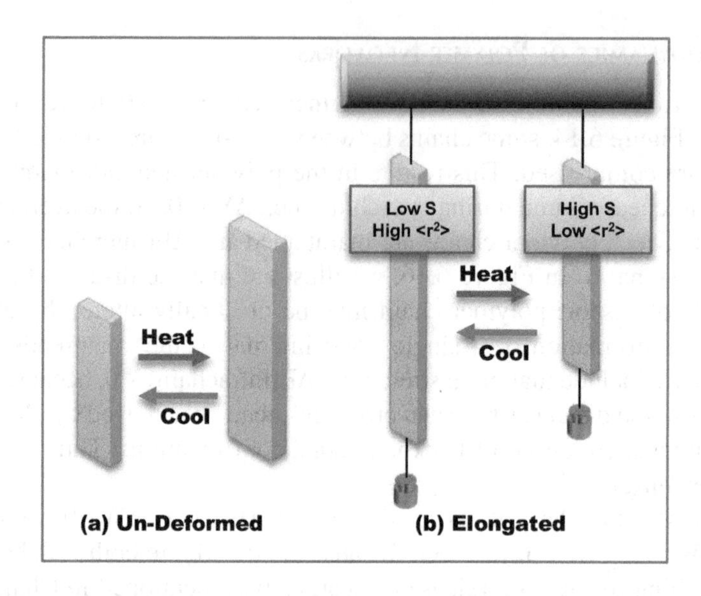

FIGURE 6.17 Length changes in unstrained (a) and strained (b) cross-linked rubber samples upon heating and cooling. In (b), the deforming force is suggested by the weight hanging from the rubber band. As T increases, the unweighted (unstressed) network will expand as most materials do, while the weighted network experiencing a force contracts. (Reprinted with permission from Shen, J. and Tonelli, A.E., *J. Chem. Educ.*, 94, 1738–1745, 2017. Copyright 2017 American Chemical Society.)

Polymer Network Topology

In an elastic network, cross-links tie together macromolecules so they may not move independently, resulting in a group of chains which bear the stress, i.e., have increased free energy when the network is strained, generally due to the reduced conformational entropy of the individual deformed chains connecting cross-links. In a polymer network cross-linked after polymerization, such as in the vulcanized natural rubbers, two types of network defects are produced: (i) polymer chain-ends and (ii) unattached loops formed by cross-links between repeat units belonging to the same polymer chain. Each is shown in Figure 6.18. They do not contribute to the elasticity of the network, aside from removing a number of potential elastically responding repeat units, because they are able to relax when strained, and so do not contribute to the equilibrium network elasticity.

Chain-ends in polymer networks were first described and discussed by Flory, and he also alluded to intramolecular loops (Flory 1944). Because they are tied to the network by only one cross-link, both defects are elastically ineffective, and when appreciable, they make estimates of network modulus very difficult. We made an effort to characterize these defects in terms of their frequency of occurrence and how much of the polymer

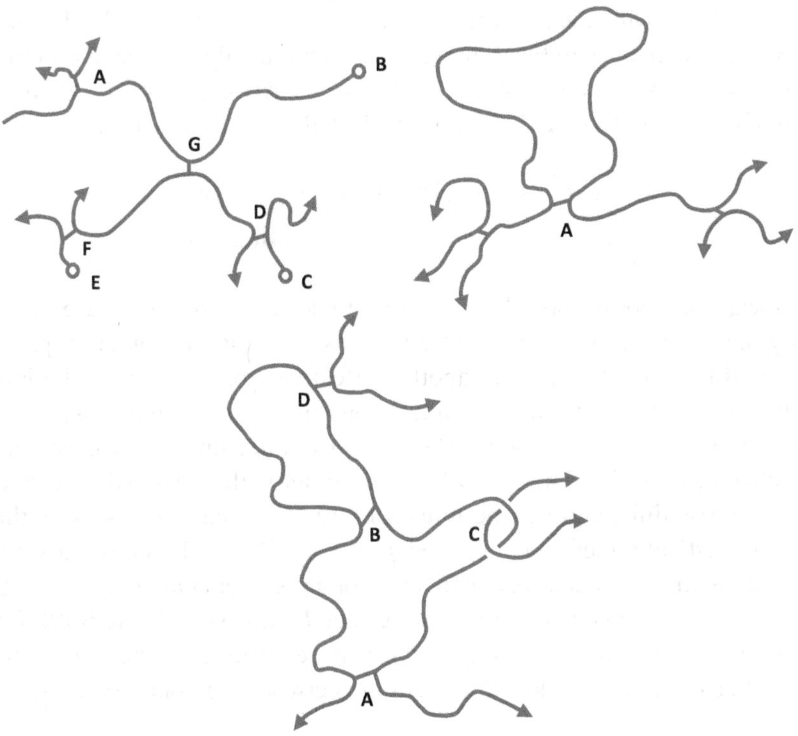

FIGURE 6.18 Upper left, free chain-ends at B, C, and E produce defects B-G, C-D, and E-F. Arrows indicate chains which are further connected to the network. Upper right, the intrachain cross-link at (A) eliminates any elastic response from the loop it forms. Bottom showing how intrachain cross-links at (A) and (B) may be elastically incorporated into the network. (Helfand, E. and Tonelli, A.E., *Macromolecules*, 7, 832, 1974; Tonelli, A.E. and Helfand, E., *Macromolecules*, 7, 59, 1974.)

network chains were elastically wasted as a result of chain-ends and intramolecular loops (Helfand and Tonelli 1974; Tonelli and Helfand 1974). We also considered how some intrachain cross-links might be elastically incorporated into the network, by means of cross-links formed between them and or by chain entanglement with elastically effective network chains. As shown in the bottom of Figure 6.18, we see that some intramolecular loops can be rendered elastically effective.

We assumed our rubber sample before cross-linking contained N polymer chains per unit volume, each with a degree of polymerization n, giving a density of $\rho = Nn$ repeat units per unit volume. After cross-linking, the density of repeat units involved in cross-links is ν (i.e., the density of cross-links is $\frac{1}{2}\nu$), or the fraction of repeat units cross-linked is $\alpha = \nu / \rho$, with a fraction g of the cross-links forming intramolecular loops. Of course the remaining fraction of cross-links $(1 - g)$ tie the network together, but a fraction of these, s, are sterile or ineffective cross-links because they join a given chain to another chain, which is not further connected to the network. The formula for s is essentially due to Flory, however, we additionally accounted for intramolecular cross-links that cannot contribute to network elasticity by inclusion of the $(1 - g)$ term, although we neglected multi-chain loops.

$$\mathbf{s} = \begin{cases} \mathbf{exp}\left[-\alpha(1-\mathbf{g})\mathbf{n}(1-\mathbf{s})\right] & \alpha(1-\mathbf{g})\mathbf{n} > 1 \\ 1 & \mathbf{otherwise} \end{cases}.$$

As mentioned above, not all of the intramolecular cross-links are elastically ineffective. Some intrachain cross-links and their resultant loops are rendered functional by having another effective cross-link along the loop (Figure 6.18). As a consequence, a fraction f of the cross-links close elastically ineffective loops, while q is the fraction of units involved in elastically effective cross-links. The sum of: (i) the units with non-sterile intermolecular cross-links and (ii) the units with intramolecular cross-links that are not elastically ineffective defines q, so $q = \alpha(1-g)(1-s)+\alpha(g-f)$.

A cross-link can occur between a unit of a polymer chain that we designate as the origin and another unit a distance L away (see Figure 6.19). L is a measure of the cross-link length and is necessarily characteristic of the type of cross-linker employed to form the network. To obtain quantitative

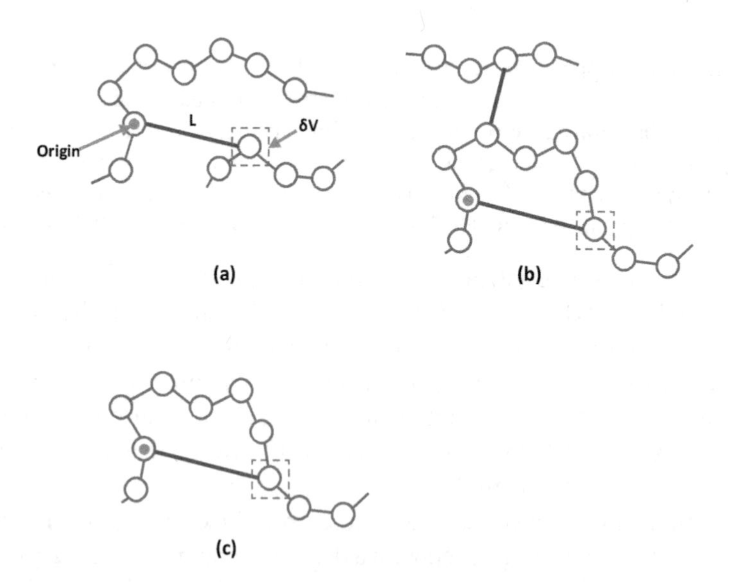

FIGURE 6.19 Cross-link between a repeat unit at the origin and a repeat unit in a volume element δV a distance L away: (a) interchain cross-link, (b) intramolecular loop rendered elastically effective by being further connected to the network, and (c) elastically ineffective intramolecular loop. (Helfand, E. and Tonelli, A.E., *Macromolecules*, 7, 832, 1974; Tonelli, A.E. and Helfand, E., *Macromolecules*, 7, 59, 1974.)

estimates of the fraction of cross-links and repeat units wasted through the formation of intrachain cross-linked loops, we need to consider several densities of units in a volume element δV a distance $\boldsymbol{L}$ from the unit located at the origin. We call ρ_1 the density in δV of monomer units belonging to the same polymer chain as the one at the origin (Figures 6.19b and c). There is a smaller density, ρ_2, of units in δV, which are part of the same macromolecule as the unit at the origin, but the intrachain loops they form have no intervening elastically effective cross-links (Figure 6.19c).

The total density (Figures 6.19a–c), which we call ρ_3, is greater than the bulk density ρ_o due to the units which are part of the polymer chain passing through the origin. An estimate for ρ_3, which neglects excluded volume effects except for not allowing densities greater than the bulk density $\boldsymbol{\rho_o}$, is $\rho_3 = \rho + \rho_1$, when $\rho + \rho_1 < \rho_0$, otherwise $\rho_3 = \rho_o$. The fraction of cross-links forming loops is $\boldsymbol{g} = \rho_1/\rho_3$, while the fraction forming elastically ineffective loops is $\boldsymbol{f} = \rho_2/\rho_3$.

To carry out the evaluation of $\boldsymbol{\rho_1}$ and $\boldsymbol{\rho_2}$, we utilize the probability density, ***W(L,k)***, that two monomer units, k units apart, on the same polymer chain are a distance *L* apart in space. ***W(L,k)*** appropriate for long polymer chains, where $\boldsymbol{k}$ = x is large, is given by the well known Gaussian function (Flory 1969).

$$W_G(L, X) = \left(\frac{3}{2\pi\langle r_x^2\rangle_0}\right)^{3/2} exp\left(-\frac{3L^2}{2\langle r_x^2\rangle_0}\right)$$

For repeat units separated by smaller values of $\boldsymbol{k}$ = x, a fourth moment correction to the Gaussian can be employed (Flory 1969).

$$W_4(L,x) \equiv \left(\frac{3}{2\pi\langle r_x^2\rangle_0}\right)^{3/2} exp\left(-\frac{3L^2}{2\langle r_x^2\rangle_0}\right)$$

$$\times\left[1-\frac{1}{8}\left(1-\frac{3}{5}\frac{\langle r_x^4\rangle_0}{\left(\langle r_x^2\rangle_0\right)^2}\right)\left(15-\frac{30L^2}{\langle r_x^2\rangle_0}+\frac{9L^4}{\left(\langle r_x^2\rangle_0\right)^2}\right)\right]$$

Because terms corresponding to small $\boldsymbol{k}$ = x<4 contribute significantly to $\boldsymbol{\rho_1}$ and $\boldsymbol{\rho_2}$, a direct enumeration of all RIS conformations was conducted. Then the probability that the end-to-end distance of a chain of $\boldsymbol{k}$ = x < 4 units was within a distance $\boldsymbol{L} \pm 0.1$ Å was evaluated.

In terms of ***W(L,k)***, $\boldsymbol{\rho_1}$ and $\boldsymbol{\rho_2}$ are given by:

$$\rho_1 = 2\sum_{k=1}^{n}\left[1-(k/n)\right]W(L,k)$$

and

$$\rho_2 = 2\sum_{k=1}^{n}\left[1-(k/n)\right]W(L,k)(1-q)^{k-1}.$$

The multiplier $[\mathbf{1}-(\boldsymbol{k}/\boldsymbol{n})]$ is a chain-end correction (Scanlan 1960). To calculate $\boldsymbol{\rho}_2$, we must include a factor $(\mathbf{1}-\boldsymbol{q})^{k-1}$ representing the probability that none of the other repeat units in the loop are effectively cross-linked to the network. The average length (α_L in repeat units) of an elastically ineffective loop may be written as:

$$\alpha_L = \frac{2}{\rho_2}\sum_{k=1}^{n}k\left[1-(k/n)\right]W(L,k)(1-q)^{k-1}.$$

Though loops on loops are double-counted, the fraction of rubber (repeat units) in such loops, $\boldsymbol{F}_L$, is a slight overestimate and is approximately $\boldsymbol{F}_L = \mathbf{1/2}(\alpha f \alpha_L)$ (Helfand and Tonelli 1974).

$\boldsymbol{\rho}_1$ and $\boldsymbol{\rho}_2$ are dominated by short loops, because short polymer chains have a greater probability of returning to the neighborhood of the origin (see Figure 6.19). In addition, longer chains/loops have an increased chance of containing elastically effective cross-links along their length. The added factor of $\boldsymbol{k}$ in the expression for $\boldsymbol{a}_{\mathrm{L}}$, however, has a profound effect and increases the contributions made by the longer loops.

Regarding entanglement of the intrachain loops formed, the degree to which entanglements act as elastically effective cross-links will determine how effective they are in restoring some of the intramolecular loops to the rolls of the elastically effective. The shorter loops are more difficult to entangle, so the larger loops are more likely to be entangled and thus connected to the elastic network (Tonelli and Helfand 1974). If a polymer is cross-linked in solution, where formation of intrachain loops is most likely, chain entanglements are less likely.

To estimate the fraction of polymer chains which are contained in elastically ineffective ends, we assumed a collection of monodisperse chains with a degree of polymerization $\boldsymbol{n}$. Assuming no intramolecular looping, we called the fraction of units which contain cross-links to other chains, which themselves are further connected by at least one more cross-link to the network, $\boldsymbol{a}'$. After a somewhat elaborate analysis, Helfand and Tonelli have shown that $a' = q = \alpha[(1-g)(1-s)+(g-f)]$, because either an elastically effective intermolecular cross-link or an intramolecular cross-link which has an elastically effective cross-link along the loop can terminate an end. They also demonstrated that the average length $\boldsymbol{\alpha}_{\mathrm{E}}$ per molecule of elastically ineffective ends, including chains unattached to the network, and $\boldsymbol{n}'$, the effective degree of polymerization accounting for looping are, respectively,

$$\alpha_E = (1/\alpha')\left[2 + (\alpha' n' - 1)exp(-\alpha' n')\right]$$

and

$$n' = n(1 - F_L).$$

Estimates of elastically ineffective cross-links and polymer segments were performed on *cis*-1,4-polyisoprene (PIP) natural rubber chains of 1,000 repeat units. An average cross-link length of 6Å was assumed; degrees of curing between 0.375% and 2% were examined; and cross-linking was assumed to occur in the dry bulk state, and in solutions with volume fractions of rubber, φ_r, of 0.5 and 0.1 (Flory 1953; Bateman 1963; Tonelli 1974c). Table 6.8 presents quantitative results for a variety of αs and φ_rs.

TABLE 6.8

Estimates of Elastically Ineffective Loops and Ends in Natural Rubber Networks

φ_r[g]	α[f], %	f[a]	α_L[b]	F_L[c]	F_E[d]	F_T[e]
1.0	0.375	0.266	26.7	0.013	0.646	0.659
	0.500	0.260	23.9	0.015	0.512	0.528
	0.625	0.254	21.8	0.017	0.419	0.436
	0.750	0.249	20.1	0.019	0.351	0.370
	0.875	0.244	18.8	0.020	0.301	0.321
	1.000	0.240	17.7	0.021	0.263	0.284
	1.125	0.236	16.8	0.022	0.233	0.255
	1.250	0.233	15.9	0.023	0.208	0.232
	1.375	0.229	15.2	0.024	0.189	0.213
	1.500	0.226	14.6	0.025	0.172	0.197
	1.625	0.223	14.0	0.025	0.158	0.184
	1.750	0.220	13.5	0.026	0.147	0.173
	1.875	0.217	13.0	0.027	0.136	0.163
	2.000	0.214	12.6	0.027	0.127	0.154
0.5	0.375	0.314	27.9	0.018	0.695	0.713
	0.500	0.333	24.9	0.021	0.557	0.578
	0.625	0.325	22.7	0.023	0.458	0.481
	0.750	0.319	21.0	0.025	0.385	0.410
	0.875	0.313	19.6	0.027	0.330	0.357
	1.000	0.308	18.4	0.028	0.288	0.317
	1.125	0.303	17.5	0.030	0.255	0.285
	1.250	0.299	16.6	0.031	0.228	0.259

(Continued)

TABLE 6.8 (*Continued*)
Estimates of Elastically Ineffective Loops and Ends in Natural Rubber Networks

φ_r[g]	α[f], %	f[a]	α_L[b]	F_L[c]	F_E[d]	F_T[e]
	1.375	0.295	15.9	0.032	0.206	0.238
	1.500	0.291	15.2	0.033	0.188	0.221
	1.625	0.287	14.6	0.034	0.173	0.207
	1.750	0.283	14.1	0.035	0.159	0.194
	1.875	0.280	13.6	0.036	0.148	0.184
	2.000	0.276	13.1	0.036	0.138	0.174
0.1	0.375	Below gel point				
	0.500	0.726	34.9	0.063	0.898	0.961
	0.625	0.712	31.7	0.070	0.811	0.881
	0.750	0.699	29.2	0.076	0.722	0.798
	0.875	0.689	27.1	0.082	0.641	0.722
	1.000	0.679	25.5	0.086	0.569	0.656
	1.125	0.670	24.0	0.091	0.508	0.599
	1.250	0.662	22.8	0.094	0.455	0.550
	1.375	0.654	21.7	0.097	0.410	0.508
	1.500	0.646	20.7	0.100	0.371	0.472
	1.625	0.639	19.9	0.103	0.338	0.441
	1.750	0.632	19.1	0.105	0.309	0.414
	1.875	0.625	18.4	0.108	0.284	0.391
	2.000	0.619	17.7	0.109	0.262	0.371

[a] f is the fraction of cross-links closing elastically ineffective loops.
[b] α_L is the average number of monomer units in an elastically ineffective loop.
[c] F_L is the fraction of polymer in elastically ineffective loops.
[d] F_E is the fraction of polymer in ends (including sol).
[e] F_T is the total fraction of defect polymer.
[f] α is the fraction of units involved in cross-links.
[g] φ_r is the volume fraction of polymer in the system being cured.

To get a sense of and some appreciation for these results, we focus only on a few. For 2% bulk dry cross-linking of PIP, the predicted closure of elastically ineffective loops wastes over 21% of cross-links, with the average loop containing approximately 13 monomer units, considerably less than the mean number (1/***a***) ~50 monomer between cross-links. This means that only about 3% of the PIP repeat units are contained in these loops. However, the amount of polymer in elastically ineffective ends is 12.7%, as a result of the wastage of cross-links, rather than the 10% predicted by ignoring intramolecular loops.

When cross-linking is carried out in solution, the extent of elastically ineffective looping sharply increases. If the same **2%** cure is done in a $\varphi_r = 0.1$ polymer solution, for example, then $f = \mathbf{0.62}$, or 62% of the cross-links are predicted to close elastically ineffective loops. These elastically ineffective loops have an average length of about 18 repeat units and constitute 11% of the PIP repeat units. In addition, 26% of the rubber would be wasted in ends, twice that wasted in the unattached ends of the dry bulk cross-linked PIP rubber.

Another profound result of intramolecular looping that is not reflected in the results in Table 6.8 is their effect on the gel point or the minimum degree of cross-linking needed to form a network incorporating all of the initial PIP chains. Ignoring loops yields a gel point or degree of cross-linking in a $\varphi_r = 0.1$ solution that is only 1/4th that required when intramolecular loops are correctly discounted. The effect of loop formation on network gelation is further expanded on below.

In general terms, we may summarize these predicted polymer network topological results by saying: (i) the number of cross-links closing elastically ineffective intramolecular loops is a small, but not negligible fraction for dry bulk curing, while quite a large fraction of cross-links are predicted to be wasted when cross-linking in solution containing only 10% rubber; (ii) the fraction of polymer repeat units contained in the intrachain loops is quite small except for the case of solution curing; and (iii) the most notable effect of the wastage of cross-links in loop formation is the resulting amount of polymer contained in chain-ends.

We had previously speculated that the elastically ineffective portions of polymer chains in a network might act like a diluent and decrease deviations from ideal elastic behavior, typically cited as empirical Mooney-Rivlin effects, even after all solvent has been removed (Mooney 1940, 1948; Rivlin 1948a, 1948b; Flory 1953; Tonelli and Helfand 1974). The calculated results on cross-linked PIP shown in Table 6.8 clearly indicate that large amounts of incompletely connected and unstrained rubber are present and could potentially serve in this manner.

Are there means for experimentally determining the types and amounts of elastically ineffective defects in polymer networks? The short answer is largely NO, but two notable attempts have been made. In the first attempt, Bica cross-linked polybutadienes (PBDs) with Molecular weights (MWs) of 60,000, 130,000, and 180,000 in solution with two different cross-linkers (Bica 1993): 1,6-hexane-bis-1,2,4-triazoline-3,5-dione and 4,4′-methylene-bis-(1,4-phenylene)di-1,2,4-triazoline-3,5-dione, abbreviated, respectively, as HMTD and MPTD and shown in Figure 6.20.

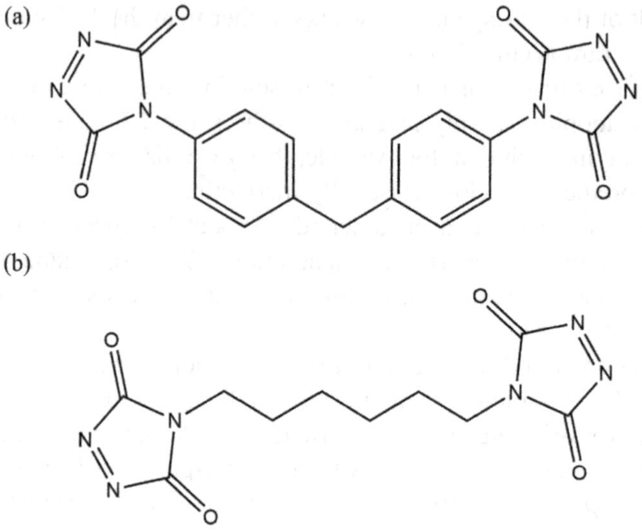

(a)

(b)

FIGURE 6.20 (a) Rigid MPTD and (b) flexible HMTD cross-linkers.

Their cross-linking reaction is illustrated in Figure 6.21, where the longer rigid cross-linker MPTD is used as the cross-linking agent. Each utilizes two 1,2,4-tri-azoline-3,5-diones (see below) as their active portions.

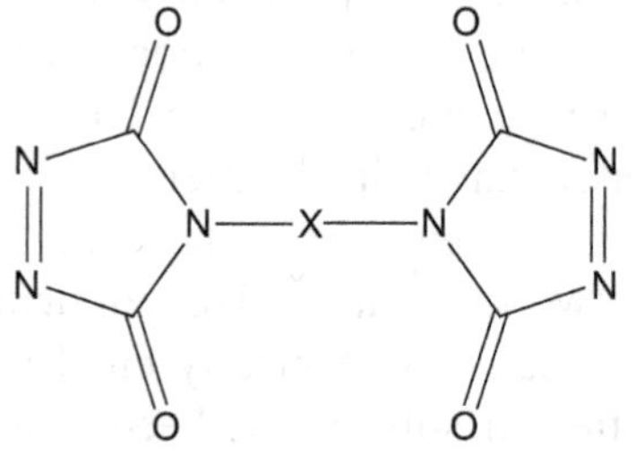

Both cross-linkers were gradually added to solutions of PBD until a cross-linked gel was formed, as monitored with viscosity measurements. Gel points were identified as the concentration of cross-linker that caused the solution viscosity to become "infinite." Bica observed that very different amounts of the two cross-linking agents were required for PBD gelation, and that both were well above the amount theoretically required (Bica 1993).

^{1}H-NMR confirmed a predominant 1,4-monomer enchainment in Bica's PBD samples, but no estimation of *cis* and *trans* double bond content was presented. We considered both all *trans* and all *cis*-1,4-PBDs in our loop formation estimates (Aminuddin et al. 1995). In addition, direct enumeration of all RIS generated conformations were used to determine *W(L, x)* for x < 4. Double bonds were assumed to be *trans* or, for the four repeat-unit

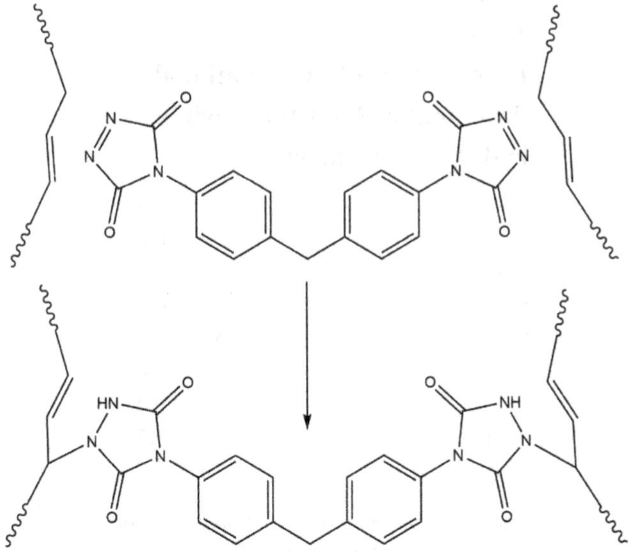

FIGURE 6.21 Cross-linking of polybutadiene with the rigid MPTD crosslinker *via* the "double en-reaction."

PBD fragments, the 1, 2, 4, and 5 unit double bonds *trans* with a *cis* double bond for the 3rd repeat unit. The second and fourth moments of the end-to-end distances of PBD chains of x repeat units, needed to determine the probability densities *W(L,x)* (see discussion above), were obtained using the RIS models developed for 1,4-PBDs by Abe and Flory (Abe and Flory 1971). We considered PBD chains well within the molecular weight range of the PBDs employed in Bica's experiments, i.e., n + 1 = 1,500 repeat units or MW = 80,000 (Bica 1993). Again, in an attempt to approximate the conditions of the experimental studies, 2% of the butadiene repeat units were assumed to be cross-linked.

The rigid MPTD cross-linker was estimated to have a cross-link length ***L*** = 13–14 Å, based on the geometries of the 1,2,4-triazoline-3,5-dione rings and the 4,4′-methylene-bis-(1,4-phenylene) fragment. This received corroborating support from an energy-optimized MPTD structure that we generated using the MM2 force field implemented by Computer-Aided Chemistry & Biochemistry(CAChe) Molecular Orbital Package (MOPAC) 6.0.

The flexible cross-linker HMTD, on the other hand, was treated as an n-octane chain with terminal bonds that are 3.55 Å in length based on the sum of the N-CH bond length and N to N = N across the ring distance for the end groups of HPTD. To calculate the end-to-end distances and probabilities of all 3^5 = 243 HMTD conformations, the RIS model appropriate to n-alkanes and polyethylene developed by Abe et al. was utilized, and

TABLE 6.9
Lengths *L* and Conformational Probabilities *P* for the Flexible HMTD Cross-Linker

L, Å	*P*
6–7	0.032
7–8	0.026
8–9	0.240
9–10	0.345
10–11	0.077
11–12	0.074
12–12.6	0.206
<*L*> = 9.92	1.0

Source: Aminuddin, A. et al., *Comput. Theor. Polym. Sci.*, 5, 165, 1995.

the results are presented in Table 6.9 (Abe et al. 1966). A mean end-to-end cross-link distance of 9.92 Å, along with cross-link lengths and their associated conformational probabilities/populations, are given there.

Two approaches were followed to calculate the fraction of cross-links wasted for the flexible HMTD cross-linker. In the first, the implicit assumption that the mobility of the cross-linker exceeds the rate of the cross-linking reaction was made, so ***L*** was assigned the average end-to-end distance of 9.92 Å. Separate *f*s were calculated for all cross-link lengths ***L*** listed in Table 6.10 in the second approach, and an average ***f*** was obtained from the probabilities associated with finding HMTD in the conformations appropriate to each ***L***. This method assumes that the cross-linking reaction is faster than the time required for HMTD to interconvert amongst all of its 243 conformations. The bulk density of $\rho = 0.925$ moles of PBD repeat units per liter of solution (6 vol%) was used, compared to the density of neat bulk PBD (ρ = 16.5 moles of repeat units per liter) (Stadler et al. 1986; Bica 1993).

From the comparisons presented in Table 6.10, we can see that experimentally, the fraction of cross-links wasted ***f*** in the formation of PBD networks in 6 vol% solutions were 60%–90% for the rigid (MPTD) and flexible (HMTD) cross-linkers, respectively. This compares favorably to the ca. 50%–70% estimated wasted cross-links. The expected fraction ***f*** of cross-links wasted for a PBD network cross-linked in the neat bulk are much reduced to 6 and 22 vol%, as calculated for the rigid (MPTD) and flexible (HMTD) cross-linkers. However, when employing cross-links of

TABLE 6.10
Observed (Bica 1993) and Estimated Cross-Link Wastage *f*

Crosslink	Estimated	Observed
MPTD (Rigid)	0.55[a] (0.42)[b]	0.60
HMTD (Flexible)		0.92
L = 6.5 Å	0.87 (0.84)	"
7.5 Å	0.86 (0.86)	"
8.5 Å	0.73 (0.76)	"
9.5 Å	0.76 (0.76)	"
10.5 Å	0.69 (0.65)	"
11.5 Å	0.63 (0.67)	"
12.5 Å	0.50 (0.57)	"
Conformationally Weighted Average	0.69 (0.71)	"

Source: Aminuddin, A. et al., *Comput. Theor. Polym. Sci.*, 5, 165, 1995.
[a] All butadiene units are assumed *trans*-1,4.
[b] All butadiene units except the 1,2,4, and 5 units in the X = 4 repeat unit fragment are assumed *cis*-1,4.

length ***L*** < 10 Å, at least 20% are expected to be elastically ineffective and therefore wasted. This leads to the following **Golden Rule:**

> ****The smallest wastage of cross-links can clearly be achieved in those networks formed in the bulk with long cross-linkers.****

The estimated amounts of wasted flexible HMTD cross-links do not depend on whether the conformationally averaged cross-link length L = 9.92 Å is assumed or if the *f*s calculated for each potential HMTD cross-link length ***L*** are averaged over all possible conformations corresponding to those ***L***s. If the time required for conformational averaging is less than the time required to form the cross-link, the estimated wastage of flexible cross-links can be more simply calculated, i.e., from the mean length of the cross-link obtained by averaging over all flexible cross-link conformations.

Our predicted wastage of cross-links underestimates those observed by a relatively minor 15%–20% and may be due to several sources: (i) neglect of the small 6% content of 1,2-butadiene units; (ii) unknown amounts of *cis*- and *trans*-1,4-butadiene units and their distributions; and (iii) our approximate treatment of the conformational characteristics of cross-linked butadiene units. It is apparent from Figure 6.21 that the attachment of one end of either cross-linker to a butadiene unit changes the structural character of that butadiene unit and consequently its conformational preferences. In addition to the presence of the bulky cross-linker itself, the double bond in the cross-linked PBD unit is shifted. Now the double bonds of the immediately adjacent butadiene units are separated by two and four single bonds from the double bond of the cross-linked unit. To the contrary, only the conformations of normal uncross-linked 1,4-butadiene units, whose double bonds are all separated by three intervening single bonds, were considered in our estimates of wasted cross-links.

Both of these structural alterations introduced by attachment of cross-links would be expected to influence the conformational characteristics of the cross-linked butadiene unit, as well as possibly those of its immediate neighbors. Because the wastage of cross-links is greatest in forming short loops ($x \leq 4$) attached to the network at only single points, we can expect our estimates of cross-link wastage to be affected by the detailed conformational characteristics of the cross-linked PBD units.

It seems apparent, that despite the relatively small differences between observed and estimated levels of wasted rigid MPTD and flexible HMTD cross-links in solution cross-linked PBD networks, the level of wasted cross-links is substantial and needs to be considered when attempting to connect/relate the physical properties and topologies of rubber and other elastic polymer networks. Thus, the Tonelli-Helfand conformational approach to estimating the fraction of network cross-links that are elastically ineffective provides a relatively facile method for gauging their level and also estimating the amount of polymer in the network which can actually support an elastic strain (Helfand and Tonelli 1974; Tonelli and Helfand 1974).

A second set of very recent studies (Zhou et al. 2012, 2014; Wang et al. 2016; Zhong et al. 2016) reported a novel experimental method for determining the number of primary loops formed in a cross-linked polymer network (Zhou et al. 2012, 2014). To determine the fraction of loops in a polymer gel, these researchers designed a network from monomers that could be carefully and precisely broken apart to dissect its loop contents. They used two monomers, one a linear oligomer with a reactive group at each end and the other a trifunctional branched monomer containing end groups reactive with the linear oligomer ends (see Figure 6.22). The linear polymer was designed to contain a strategically placed, readily cleavable chemical bond that produces a short and a long fragment upon cleavage. Measurement of network disassembly products allowed a measure of the number of primary loops and their wasted cross-links contained in their networks (see Figure 6.23). Even more recently, they

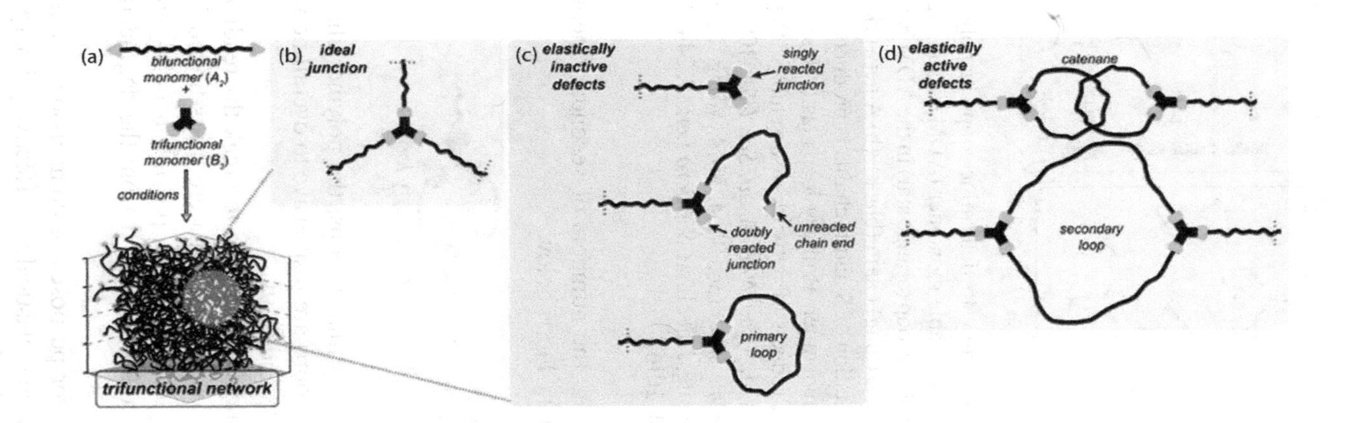

FIGURE 6.22 Schematic of trifunctional end-linked network formation. (a) Reaction between a bifunctional A2 monomer and a trifunctional B3 monomer to form a network. (b) Triply reacted "ideal" junctions. (c) Elastically inactive molecular defects. (d) Two possible elastically active defects. (Reprinted with permission from Zhou et al., *J. Am. Chem. Soc.*, 136, 26, 9464–9470, 2014. Copyright 2014 American Chemical Society.)

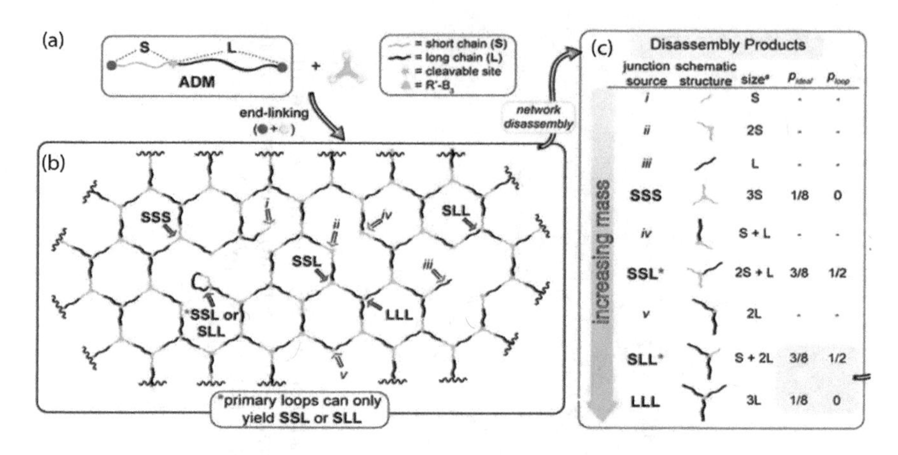

FIGURE 6.23 Network disassembly to determine loops in polymer gels. (a) An end-linkable asymmetric degradable monomer (ADM). After cleavage of an asymmetrically placed degradable group (orange star) in the ADM backbone leads to a S (short) chain (blue) and a L (long) chain (black). (b) A network with unique junctions, in terms of the orientation of S and L chains is produced by linking of the ADM and a trifunctional monomer. Primary loops (asterisks) cannot reside at SSS or LLL junctions. (c) Products of network disassembly. (Reprinted with permission from Zhou, H.X. et al., *Proc. Natl. Acad. Sci. USA*, 109, 19119, 2012; Zhou, H.X. et al., *J. Am. Chem. Soc.*, 136, 9464, 2014; Wang, R. et al., *Accts. Chem. Res.*, 49, 2786, 2016; Wang, J. et al., *ACS Macro Lett.*, 7, 244, 2018; Zhong, M.J. et al., *Science*, 353, 1264, 2016.)

developed means to estimate the numbers of secondary network loops that are compared to primary loops below.

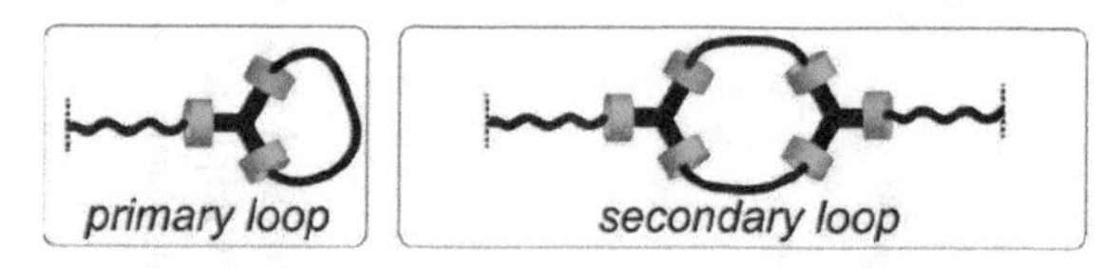

Unlike elastically ineffective primary loops, the polymer chain portions which constitute secondary loops are responsive to overall network deformation (also see Figures 6.18 and 6.19).

Careful dissection techniques were used to break the readily cleavable bonds and were applied to networks that were cross-linked in solution. Cleavage of the networks yielded only six possible network junctions, of which, only two products could be formed from the degradation of primary loops.

Unfortunately, however, for purposes of comparison to our predicted network defects (chain-ends and elastically ineffective primary loops), the results of their clever investigations of cleavable networks are not completely relevant. The linear end-functionalized monomers they used were short, i.e., 16–32 ethylene-glycol repeat units, and limit the relevance of comparing the numbers of elastically ineffective primary loops found in their experiments to those we predicted and that have been corroborated on solution cross-linked PBD networks (Bica 1993; Helfand and Tonelli 1974; Tonelli and Helfand 1974; Aminuddin et al. 1995; Zhou et al. 2012, 2014; Wang et al. 2016; 2018; Zhong et al. 2016).

Having demonstrated that it is possible to characterize the topologies of cross-linked polymer networks by connecting them to the microstructural-dependent conformational characteristic of their constituent chains, we now describe an effort to estimate the modulus of a defect free network made by end-linking dimethyl siloxane oligomers.

Modulus of a Polymer Network

To be relevant and useful, a molecular theory of rubber-like elasticity has to predict the elastic modulus of a polymer network based solely on a knowledge of network topology and the conformational characteristics of the constituent deformed network chains between cross-links. To accomplish this goal, the relationship between the externally applied macroscopic network deformation and the molecular response of the network chains must be established, which in the current reality must be presumed. We need to know if the network junctions or cross-links affinely translate and faithfully track the macroscopic deformation. If they do not, do they instead fluctuate over time independently from the macroscopic state of deformation, without any hindering interference from the presence of neighboring chains. If this is the case, then all chains are assumed to be "phantom" in nature and able to pass through each other when sampling their conformations in both their at rest and stretched states.

A middle ground between the affine and phantom network models mentioned above has been suggested, which permits fluctuation of network junctions, but which are constrained by the interpenetration of their attached chains with neighboring chains and junctions. This has been called the constrained junction model (James and Guth 1947; Flory 1953, 1977a, 1977b). Each of these molecular theories of rubber elasticity assume long conformationally flexible chains between cross-links, whose lengths are characterized by Gaussian statistics.

Comparison of the behavior expected from molecular theories with the elastic response observed for real elastic networks is made difficult by two factors: (1) uncertain topology of experimental networks (dangling chain-ends and unattached loops, the distribution of cross-linked chain lengths, etc.), and (2) the detailed conformational responses of the chains between junctions as the network is deformed (Helfand and Tonelli 1974; Tonelli and Helfand 1974; Aminuddin et al. 1995).

We attempted to address these two difficulties by first forming networks through the end-linking of poly(dimethyl siloxane) (PDMS) oligomers to eliminate uncertain network topologies. Because the chains between cross-links are short, these networks are tough, with low extensibility and high moduli. Also because the network chains between cross-links are short, their conformational responses may be assessed by direct enumeration of all of their RIS conformations without assumption of Gaussian chain statistics. This permits a realistic evaluation of their conformational partition functions, energies, entropies, and free energies, as a function of their end-to-end lengths (***r***), and enables evaluation of the force required to extend and compress the end-to-end lengths of deformed network chains between cross-links.

Adoption of this combined experimental and computational approach was prompted by the hope that the comparison of observed and estimated moduli would not be plagued either by uncertain network topology and/or the assumption that the conformational responses of network chains are purely Gaussian. To the contrary, we sought a direct comparison between the force needed to achieve the experimental macroscopic deformation and the calculated force for deforming individual, but rather short, PDMS oligomer chains. The latter force is dependent only upon the assumed connection between the deformations of the individual end-linked molecular chains and the macroscopic network as a whole.

Following the experimental procedure of Mark and Sullivan, we employed PDMS oligomers (5-, 6-, 9-, and 10-mers) that were terminated with hydroxyl groups. After careful drying, they were tetrafunctionally end-linked with tetraethylorthosilicate, which was added in 1:2 stoichiometric amounts of tetraethylorthosilicate ethoxy groups:PDMS hydroxyl end-groups (see Figure 6.24). We employed 1 wt% stannous-2-ethylhexanoate relative to the weight of PDMS oligomers to catalyze their end-linking reactions.

Thoroughly mixed network ingredients were poured into aluminum molds and reacted for two days at room temperature, subsequently extracted with benzene, and dried (Mark and Sullivan 1977). Approximately 4% of the total PDMS oligomer starting weight was uncross-linked and remained soluble, which is the amount of inert cyclic material typically found (2%–6%) in cross-linked PDMS samples (Wright and Beavers 1986).

Strips cut from the extracted and dried elastomer networks were used to measure stress-strain isotherms. The network sample strips were mounted between two clamps. The lower clamp remained stationary, while the upper clamp was attached to a vertically extendable force gauge. Output from the force gauge was monitored, but only after any network relaxations had ceased were equilibrium force readings recorded. The distance between two pre-inscribed lines marked on the test network strip were measured with

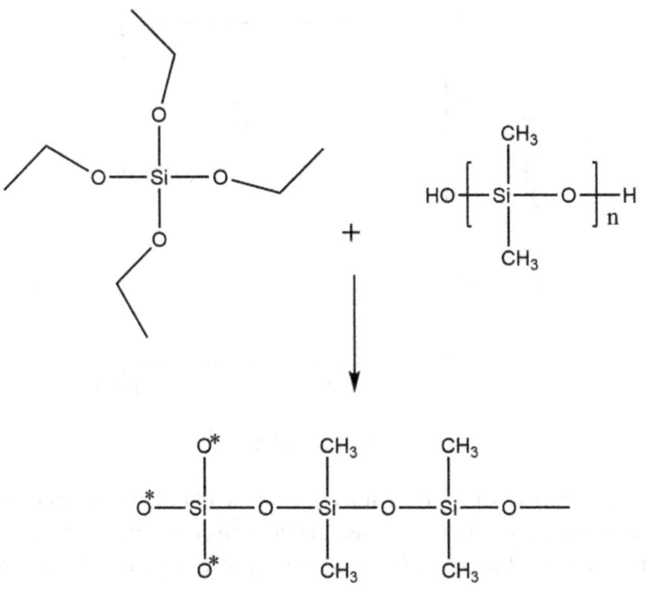

FIGURE 6.24 End-linking PDMS oligomers with tetraethylorthosilicate to form ideal loop-free networks. (Tonelli, A.E. and Andrady, A., *Comput. Theor. Polym. Sci.*, 6, 103, 1996.)

a cathetometer and taken as the strain. The stress-strain measurements were conducted with the sample maintained in a nitrogen atmosphere inside a glass vessel immersed in a constant temperature water bath. Force and strain data were recorded as the sample strain was progressively increased in a series of steps. For the PDMS 9-mer network discussed here, only one specimen of five that were tested yielded a useful stress-strain curve, while the other four sample strips failed early during extension or in the clamping process.

Using benzene as the swelling agent, swelling ratios of the networks were determined. A weighed amount of the network sample was placed in excess benzene at 25°C for about 48 hrs. The swollen sample was then directly transferred into a weighing bottle and accurately weighed. Using the known density of benzene, the volume change in the swollen sample was calculated from the gain in sample weight.

The stress-strain data for the network formed by tetrafunctionally end-linking PDMS 9-mers are presented in Figure 6.25. At break ($\lambda_{max} = 1.12$), the modulus of the end-linked 9-mer PDMS network was observed to be $\mathbf{2.2 \times 10^6\ N/m^2}$. Note that the observed and calculated (see below) moduli of these short chain PDMS networks are ca. 20 times higher than those normally observed in PDMS rubbers, where the chains between cross-links typically contain ca. 200 monomer units (Curro and Mark 1984).

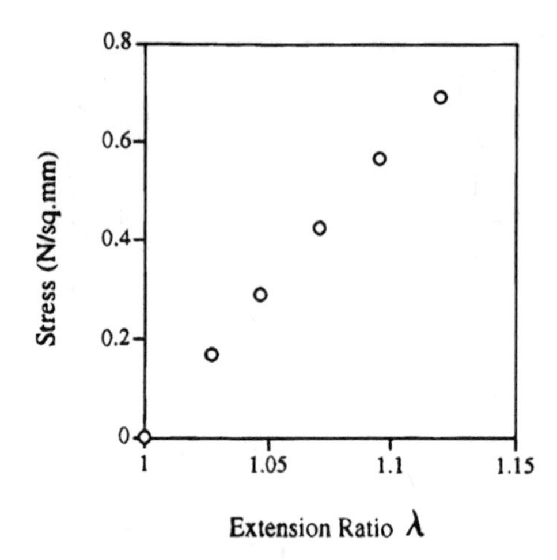

FIGURE 6.25 Stress-strain data for end-linked PDMS 9-mer network chains having a number average molecular weight between cross-links of 660. (Reprinted with permission from Tonelli, A.E. and Andrady, A., *Comput. Theor. Polym. Sci.*, 6, 103, 1996.)

The higher moduli of our model PDMS oligomer networks are the result of increases in the conformational energies, as well as larger decreases in conformational entropies, when the short chains between cross-links are stretched or compressed during network extension. These are caused by the large fraction of oligomer bonds that are forced to change conformations as the network is deformed. When the same PDMS 9-mer network is swollen to equilibrium in benzene at 25°C, it is interesting to note that V(swollen)/V(dry) = 1.56 = λ^3, leading to a λ_{max} (isotropic) = $(1.56)^{1/3}$ = 1.16. This is in very close agreement with the λ_{max} = 1.12 observed at break, strongly suggesting that the elastic behavior of the end-linked PDMS 9-mer network is isotropic and occurs with no change in network volume as assumed.

To make the connection between the macroscopic and molecular network deformations, we adopted the 3-chain network model of James and Guth, as illustrated in Figure 6.26. A cube, with sides L = $\langle r \rangle$, the average end-to-end length of an undeformed PDMS oligomer, is assumed to contain three oligomer chains, each perpendicular to the three pairs of parallel sides of the cube. Because, as mentioned previously, the deformation of an elastic network generally occurs without a change in volume, and assuming affine network deformation, extension of the at rest cube transforms it into a parallelepiped of length $\lambda\langle r \rangle$ and cross-section of $(\langle r \rangle/\lambda^{1/2})^2$ with the same volume $\langle r \rangle^3$ as the unstretched cube. Upon stretching, one PDMS oligomer chain is extended from $\langle r \rangle$ to $\lambda_{max}\langle r \rangle = r_{max}$, while the other two chains are compressed from $\langle r \rangle$ at rest to $d = (\langle r \rangle^3/r_{max})^{1/2}$. Such a 3-chain oligomer network has a density $\rho = 3\ M/(N_A\langle r \rangle^3)$, where

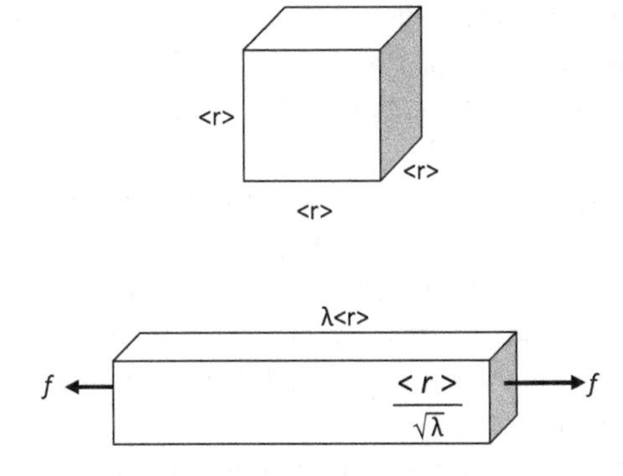

FIGURE 6.26 Three-chain network. Unstrained state (top) and simple extension (bottom). (James, H. M., Guth, E. (1943), J. Chem. Phys. 11, 455).

M is the oligomer molecular weight. The densities calculated this way for the PDMS 3-chain networks are, respectively, $\rho = 1.023$ and 0.996 g/cm^3 for the PDMS 9- and 10-mer cubes, which compare favorably to the bulk density of PDMS = 0.975 g/cm^3.

We estimate that less than 1% of the PDMS 9-mers are attached at both of their ends to the same cross-link, which would produce an elastically ineffective loop and thereby remove such chains from the population of PDMS 9-mers able to respond elastically to network deformation (Tonelli and Helfand 1974; Helfand and Tonelli 1974; Aminuddin et al. 1995). This conclusion is based on the probabilities calculated for all PDMS 9-mer conformations with end-to-end vectors whose x, y, and z components are each less than 0.2 Å from those of the remaining oxygen atoms (O*) of the cross-link.

O* CH$_3$ CH$_3$
O*—Si—O—Si—O—Si—O—
O* CH$_3$ CH$_3$

The force f required to deform each of the 3-chains in the oligomer network is given by $f = (\Delta E_{conf} - T\Delta S_{conf})/\Delta r$. Previously we suggested that the resistance to stretching an elastic polymer network is derived solely from the reduction in the conformational entropies ΔS_{conf} of the deformed chains. This is true as long as the polymer chains between cross-links are long. As we saw in Figure 6.16, when this is the case, the conformations of only a small number of the bonds connecting cross-links need to change in response to stretching or compressing the distance between cross-links and the chains between them, and in this case ΔE_{conf} is negligible. However, in our PDMS 9-mer network there are fewer than 20 Si-O bonds between cross-links, and the number or fraction of bonds that must alter their conformations to accommodate network deformations, both stretching and compression, are not negligible.

Alternatively, the force f required to deform each of the 3-chains in the oligomer network may also be estimated as $f = -(\mathbf{RT}/\mathbf{Z}_r)(d\mathbf{Z}_r/dr)$. $\mathbf{Z}_r$ is that portion of the total partition function $\mathbf{Z}$ of all PDMS oligomer conformations with end-to-end lengths $= r$, and is readily evaluated from $\mathbf{Z}_r = P_{conf(r)}\mathbf{Z}$, where $P_{conf(r)}$ is the sum of probabilities for all conformations with end-to-end lengths of r.

Consistent with the placement of each chain perpendicular to one pair of parallel cube sides in James and Guth's 3-chain network model, and for affine network deformation, one cube chain is stretched from $r = <r>$ to $r = r_{max}$ by a force of extension f_e, while the two chains perpendicular to the stretch direction are assumed to be compressed from $r = <r>$ to $r = (<r>^3/r_{max})^{1/2}$, each with a compression force f_c. Thus, the total force f_{tot} expected to resist the uniaxial extension of the 3-chain cube is $f_{tot} = f_e + 2f_c$. The modulus σ of the 3-chain tube is then obtained directly from f_{tot} and is $\sigma = f_{tot}/\{A_o[\lambda - (1/\lambda^2)]\}$, where $A_o = <r>^2$ is the cross-sectional area of the undeformed cube.

A modulus of $\mathbf{1.2 \times 10^6\ N/m^2}$ was calculated from the dependence of the conformational energies and entropies on the lengths r of the 9-mer chains (see Figure 6.27). Z_r, E_{conf}, and S_{conf} were calculated using the Flory

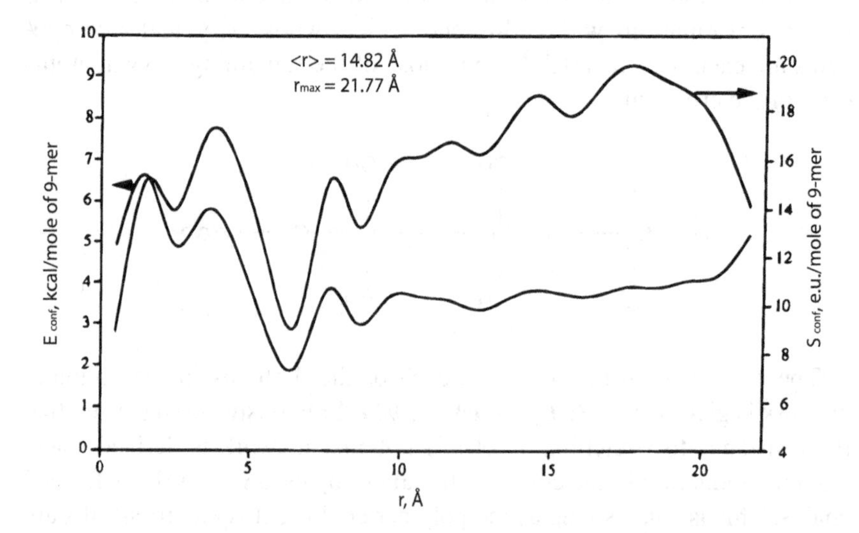

FIGURE 6.27 E_{conf} and S_{conf} calculated for a PDMS 9-mer (FCM-RIS model, Flory, P.J. et al., *J. Am. Chem. Soc.*, 86, 146, 1964.) as a function of its end-to-end length r. (Reprinted with permission from Tonelli, A.E. and Andrady, A., *Comput. Theor. Polym. Sci.*, 6, 103, 1996.)

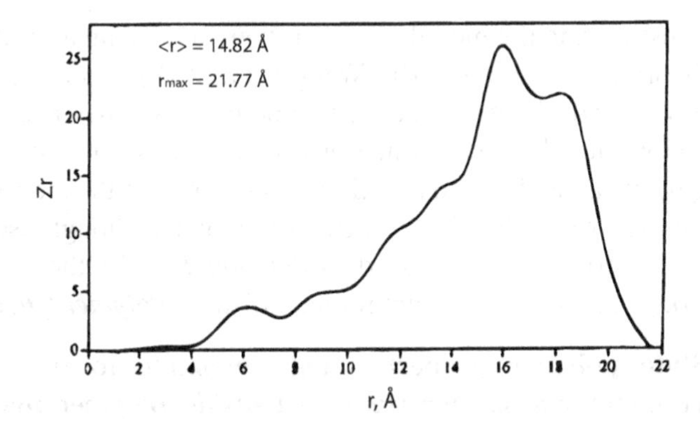

FIGURE 6.28 The partition function *Zr* of PDMS 9-mer conformations with end-to-end lengths *r* calculated as a function of *r*. (Reprinted with permission from Tonelli, A.E. and Andrady, A., *Comput. Theor. Polym. Sci.*, 6, 103, 1996.)

et al. RIS model developed for PDMS chains (Flory et al. 1964). From the dependence of $\boldsymbol{Z_r}$ on $\boldsymbol{r}$ (see Figure 6.28), a modulus of **10^7 N/m²** was obtained. These calculated moduli (**1.2 and 10 × 10^6 N/m²** nicely bracket the measured modulus of **2.2 × 10^6 N/m²**). A more recent RIS model was developed for PDMS by Bahar et al. (1991), and when similarly employed to estimate the moduli of the 9-mer network, values of **1.1** and **6.0 × 10^6 N/m²** were obtained, respectively, from the dependence of $\boldsymbol{E_{conf}}$ and $\boldsymbol{S_{conf}}$ and ***Z*** on ***r***, and so also nicely bracket the measured network modulus.

The second feature to note in Figure 6.27 is that a minimum in $\boldsymbol{E_{conf}}$ of the PDMS 9-mer appears at ***r*** = 6Å. This is a consequence of the fact that the all *trans* conformation of PDMS has the lowest energy and because the Si-O-Si valence angle (143°) is more than 30° greater than the typical tetrahedral value (Flory et al. 1964). This results in a closed cyclic structure (***r*** = 0) for the all *trans* lowest energy conformer of the PDMS 11-mer, while the lowest energy all *trans* PDMS 9-mer conformation forms an incompletely closed cycle with ***r*** = 6Å.

Our prediction of the modulus of a tough end-linked PDMS network of short 9-mer PDMS chains was based essentially through estimation of the energies (E_{conf}), entropies (S_{conf}), and lengths (*r*), for every 9-mer conformation. This permitted determination of the forces required to stretch or compress the chains as the network was deformed, once again illustrating the utility of our ***Inside*** polymer chain microstructure ←→ ***Outside*** polymer material property approach for establishing truly relevant structure ←→ property relations for polymers.

As previously mentioned, Zhou et al. (2012), (2014); Wang et al. (2016), (2018); and Zhong et al. (2016) have developed a clever experimental means to determine the numbers of primary inelastic, secondary, and higher elastic loops in end-linked networks. They have also developed a method for estimating the modulus of their networks using a "real elastic" network theory (Zhong et al. 2016; Wang et al. 2018). However, because this theory is strictly based on network topology, assumes phantom network behavior, and chain conformations with a Gaussian distribution of chain lengths, it cannot be used to estimate and compare the moduli of networks composed from chemically distinct polymers. Though useful, it is instead a microstructurally independent ***Polymer Physics*** theory of polymer networks that does not account for their distinct ***Polymer Chemistries***.

Solid-State polymer properties not amenable to the *Inside* polymer chain microstructure ←→ *Outside* polymer material property approach

As we have illustrated above with several examples, knowledge of the conformational preference of individual polymer chains that are dictated by their chemical microstructures, i.e., their ***Polymer Chemistries***, can often lead to some understanding of the behaviors of their solid materials. However, there remain many solid-state behaviors that cannot be directly connected to the conformational preferences of their constituent polymer chains. Such solid-state behaviors and properties are dominated by the interactions between polymer chains, rather than the inherent conformational preferences of their individual chains. Thus, just as the dynamic behaviors of liquid polymers are currently intractable, so are those of solid polymer materials that are dominated by the interactions between their chains.

Two such properties are the magnitudes of amorphous polymer T_gs, discussed previously, and the mechanical strengths of solid polymer materials. Whether a solid polymer is amorphous or semi-crystalline, its constituent chains or portions of them are not packed regularly. This makes estimation of the interaction energy between the chains and the configurational entropy of chain packing virtually impossible. As a consequence, when we deform the solid polymer material, we are unable to determine how much its energy and entropy are changed in opposition to the deformation, and that constitute its mechanical strength.

REFERENCES

Abe, Y., Flory, P. J. (1971), *Macromolecules*, 4, 219.

Abe, A., Jernigan, R. L., Flory, P. J. (1966), *J. Am. Chem. Soc.*, 88, 631.

Abis, L., Floridi, G., Merlo, E., Po, R., Zannoni, C. (1998), *J. Polym. Sci. Part B: Polym. Phys.*, 36, 1557, 1339–1343, 199.

Aminuddin, A., Burke, J., Eaton, P., Huang, L., Vasanthan, N., Tonelli, A. E. (1995), *Comput. Theor. Polym. Sci.*, 5, 165.

Anet, F. A. L., Ahmad, M. (1964), *J. Am. Chem. Soc.*, 86, 119.

Bahar, I., Zuniga, I., Dodge, I., Mattice, W. L. (1991), *Macromolecules*, 24, 2986.
Bateman, L., Ed. (1963), *The Chemistry and Physics of Rubber-Like Substances*, Wiley and Sons, New York, Chapters 15, 16, and 19.
Bica, C. I. D. (1993), *Euro. Polymer. J.*, 29, 1339.
Chiang, R., Flory, P. J. (1961), *J. Am. Chem. Soc.*, 83, 2857.
Cholli, A. L., Dumais, J. J., Engel, A. K., Jelinski, L. W. (1984), *Macromolecules*, 17, 2399.
Curro, J. G., Mark, J. E. (1984), *J. Chem. Phys.*, 80, 452.
DiMarzio, E. A., Gibbs, J. H. (1959), *J. Polym. Sci.*, 40, 121.
Fateley, W. G., Harris, R. K., Miller, F. A., Witkowski, R. E. (1965), *Spectrochim. Acta*, 2, 231.
Flory, P. J. (1944), *Chem. Rev.*, 35, 57.
Flory, P. J. (1953), *Principles of Polymer Chemistry*, Cornell University Press, Ithaca, NY.
Flory, P. J. (1967), *J. Am. Chem. Soc.*, 89, 1798.
Flory, P. J. (1969), *Statistical Mechanics of Chain Molecules*, Wiley-Interscience, New York.
Flory, P. J. (1977a), *J. Chem. Phys.*, 66, 5720.
Flory, P. J. (1977b), In *Contemporary Topics In Polymer Science*, Pearce, E. M.; Schaefgen, J. R., Eds., Plenum, New York, Vol. 2.
Flory, P. J., Crescenzi, V., Mark, J. E. (1964), *J. Am. Chem. Soc.*, 86, 146.
Fox, T. G., Flory, P. J. (1950), *J. Appl. Phys.*, 21, 581.
Fox, T. G. (1956), *Bull. Am. Phys. Soc.*, 1, 123.
Green, J. H. S., Kynaston, W., Gebbie, H. A. (1962), *Nature (London)*, 195, 595.
Goodyear, C. (1837), US patent no. 240.
Gordon, M., Macnab, I. A. (1953), *Trans. Faraday Soc.*, 49, 31–39.
Heijboer, J. (1968), *J. Polym. Sci., Part C*, 16, 3755.
Heijboer, J. (1969), *Brit. Polym.*, 1, 3.
Helfand, E., Tonelli, A. E. (1974), *Macromolecules*, 7, 832.
Hill, T. L. (1960), *An Introduction to Statistical Mechanics*, Addison-Wesley, Reading Massachusetts.
Hirooka, M., Kato, T. (1974), *J. Polym. Sci., Lett. Ed.*, 12, 31.
Immirzi, A., Tedesco, C., Monaco, G., Tonelli, A. E. (2005), *Macromolecules*, 38, 1223.
Itoh, T. (1976), *Japn. J. Appl. Phys.*, 15, 2295.
James, H. M., Guth, E. (1947), *J. Chem. Phys.*, 15, 669.
Johnston, N. W. (1976), *J. Macromol. Sci., Part C: Rev. Macromol. Chem.*, 14, 214.
Kotek, R., Pang, K., Schmidt, B., Tonelli, A. (2004), *J. Polym. Sci.: Part B: Polym. Phys.*, 42, 4247.
Mark, J. E., Sullivan, J. L. (1977), *J. Chem. Phys.*,66, I 006.
Mooney, M. (1940), *J. Appl. Phys.*, 11, 582.
Mooney, M. (1948), *J. Appl. Phys.*, 19, 434.
Niven, W. D. (1890), *The Scientific Papers of James Clerk Maxwell*, Cambridge University Press Cambridge, UK, pp. 713–741.
Polyakova, A., Liu, R. Y. F., Schiraldi, D. A., Hiltner, A., Baer, E. (2001), *J. Polym. Sci. Part B: Polym. Phys.*, 39, 1889.

Rivlin, R. S. (1948a), *Trans. Royal Soc.*, A240, 459, 491, 509.
Rivlin, R. S. (1948b), *Trans. Royal Soc.*, A241, 379.
Scanlan, J. (1960), *J. Polym. Sci.*, 43, 501.
Semler, J. J., Jhon, Y. K., Tonelli, A. E., Beevers, M., Krishnamoorti, R., Genzer, J. (**2007**), *Adv. Mater.*, 19, 2877.
Shen, J., Caydamli, Y., Gurarslan, A., Li, S., Tonelli, A. E. (2017), *Polymer*, 124, 235.
Shen, J., Tonelli, A. E. (2017), *J. Chem. Ed.*, 94, 1738–1745.
Shen, Z., Luo, F., Lei, X., Ji, L., Wang, K. (2016), *J. Polym. Res.*, 23, 212.
Silver, H. G., Wood, J. L. (1964), *Trans. Faraday Soc.*, 60, 5.
Slichter, W. P. (1955), *J. Appl. Phys.*, 26, 1099.
Slichter, W. P. (1959), *J. Polym. Sci.*, 35, 77.
Stadler, R., Freitas, L., Jacobi, M. (1986), *Makromol. Chem.*, 187, 723.
Tonelli, A. E. (1970), *J. Chem. Phys.*, 52, 4749.
Tonelli, A. E. (1971), *J. Chem. Phys.*, 54, 4637.
Tonelli, A. E. (1972), *Macromolecules*, 5, 558.
Tonelli, A. E. (1973a), *J. Polym. Sci., Polym. Lett., Ed.*, 11, 441.
Tonelli, A. E. (1973b), *Macromolecules*, 6, 503.
Tonelli, A. E. (1974a), *Macromolecules*, 7, 632.
Tonelli, A. E. (1974b), *Analytical Chemistry*, Vol. 3, R. S. Porter and J. F. Johnson, Eds., Plenum, New York, p. 89.
Tonelli, A. E. (1974c), *Polymer*, 15, 194.
Tonelli, A. E. (1975), *Macromolecules*, 8, 54.
Tonelli, A. E. (1977a), *Macromolecules*, 10, 633.
Tonelli, A. E. (1977b), *J. Polym. Sci., Polym. Phys. Ed.*, 15, 2051.
Tonelli, A. E. (1986), *Encyclopedia of Polymer Science and Engineering*, 2nd Ed., Wiley, New York, Vol. 4, p. 120.
Tonelli, A. E. (2001), *Polymers from the Inside Out*, Wiley-VCH, New York.
Tonelli, A. E. (2002), *J. Polym. Sci.: Part B: Polym. Phys.*, 40, 1254.
Tonelli, A. E., Andrady, A. (1996), *Comput. Theor. Polym. Sci.*, 6, 103.
Tonelli, A. E., Helfand, E. (1974), *Macromolecules*, 7, 59.
Tonelli, A. E, Jhon, Y. K., Genzer, J. (2010), *Macromolecules*, 43, 6912.
VanderHart, D. L., Bohm, G. G. A., Mochel, V. D. (1981), *Polym Prepr (Am. Chem. Soc. Div. Polym. Chem.)*, 22, 112.
Wang, R., Sing, M. K., Avery, R. K., Souza, B. S., Kim, M., Olsen, B. D. (2016), *Accts. Chem. Res.*, 49, 2786.
Wang, J., Lin, T.-S., Gu, Y., Wang, R., Olsen, B. D., Johnson, J. A. (2018), *ACS Macro Lett.*, 7, 244.
Williams, A. D., Flory, P. J. (1967), *J. Polym. Sci.*, A–2, 5, 417.
Wright, P. V., Beevers, M. S. (1986), In *Cyclic Polymers*, Semlyen, J. A., Ed., Elsevier, London, p. 85.
Yoon, D. Y., Sundararajan, P. R., Flory, P. J. (1975), *Macromolecules*, 8, 776.
Zhong, M. J., Wang, R., Kawamoto, K., Olsen, B. D., Johnson, J. A. (2016), *Science*, 353, 1264.
Zhou, H. X., Woo, J., Cok, A. M., Wang, M. Z., Olsen, B. D., Johnson, J. A. (2012), *Proc. Natl. Acad. Sci. USA.*, 109, 19119.
Zhou, H. X., Schon, E. M., Wang, M. Z., Glassman, M. J., Liu, J., Zhong, M. J., Diaz, D., Olsen, B. D., Johnson, J. A. (2014), *J. Am. Chem. Soc.*, 136, 9464.

DISCUSSION QUESTIONS

1. We illustrated the use of polymer RIS models to evaluate their conformational flexibilities as represented by their conformational partition functions Z_{conf}. This was done because the larger the conformational partition function, the more low energy conformations are available to the polymer chain.
   a. Populations of bond conformations were obtained from Z_{conf} and used to evaluate γ-gauche shielding of ^{13}C nuclei and the populations of stereo-isomeric sequences in vinyl polymers, for example. Describe how this was done.
   b. Please indicate how Z_{conf} was used to understand the comonomer-sequence-dependent T_gs of amorphous polymers and the T_ms of semi-crystalline polymers.
2. We also used Z_{conf}s calculated from polymer RIS models to determine thermodynamic quantities like E_{conf}, S_{conf}, and A_{conf} for individual polymer chains averaged over all their myriad conformations. Give examples how this was done and for what purposes they were used.
3. In addition, spatial polymer chain properties like dimensions ($<R_g^2>_o$ and $<r^2>_o$) and dipole moments ($<\mu^2>_o$) were calculated by averaging over all RIS conformations. Indicate how this was accomplished and give examples of how these were applied to understand material behaviors.
4. In particular, describe how the polymer chain properties mentioned in numbers 2 and 3 were used to understand the topologies and moduli of elastic polymer networks.
5. What is the principal reason that 1,4-linked phenyl rings in polymer backbones, such as that in PET below, essentially create a dynamically free rotation link (—) between the *para*-linked chain backbone portions preceding and succeeding the phenyl ring?

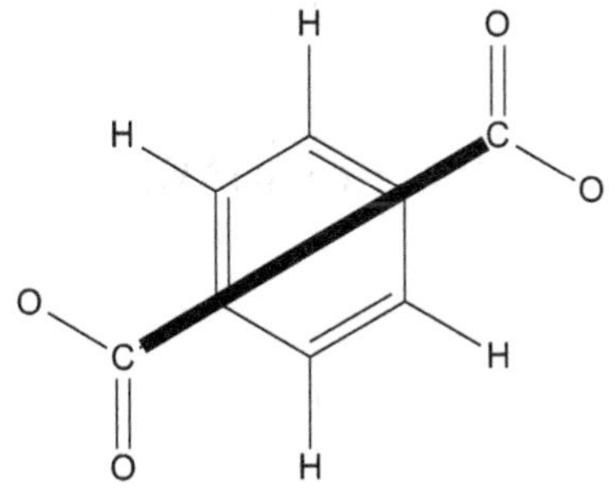

PET

6. Describe some material consequences for polymers that have dynamically freely rotating phenyl rings in their backbones.

APPENDIX 6.1

Derivation of $\Delta S_V(\Delta S_{exp}) = \Delta V_m(\partial P/\partial T)_V = (\alpha/\beta)\Delta V_m$, where α, β, and ΔV_m are the thermal expansion coefficient, the isothermal compressibility (both measured on the melt just above T_m), and the volume change observed on melting, respectively.

$$\alpha = (1/V)(\partial V/\partial T)_P \text{ and } \beta = -(1/V)(\partial V/\partial P)_T$$

$$A = E - TS,\ dA = dE - TdS - SdT = dQ - dW - TdS - SdT =$$

$$TdS - PdV - TdS - SdT = -PdV - SdT =$$

$$(\partial A/\partial V)_T\, dV + (\partial A/\partial T)_V\, dT$$

$$\partial\left[(\partial A/\partial V)_T\right]/\partial T = \partial\left[(\partial A/\partial T)_V\right]/\partial V$$

$$(\partial A/\partial V)_T = -P \text{ and } (\partial A/\partial T)_V = -S \rightarrow$$

$$-(\partial P/\partial T)_V = -(\partial S/\partial V)_T \text{ (A Maxwell Relation)}$$

Chain Rule:

$$(\partial P/\partial T)_V \times (\partial T/\partial V)_P \times (\partial V/\partial P)_T = -1$$

$$(\partial P/\partial T)_V \times (-1/\alpha V)_P \times (-\beta V)_T = -1 \text{ or}$$

$$-(\partial P/\partial T)_V \times (\beta/\alpha) = -1 \text{ or} = (\partial P/\partial T)_V = (\alpha/\beta) = (\partial S/\partial V)_T$$

Thus,

$$\Delta S_{vol} = (\alpha/\beta)\Delta V_m$$

7 Biopolymer Structures and Behaviors with Comparisons to Synthetic Polymers

INTRODUCTION

Chapters 3 and 4 demonstrated how the short-range microstructures of synthetic polymers can be determined by ^{13}C-Nuclear Magnetic Resonance (^{13}C-NMR) and how they affect local polymer chain conformations. Then in Chapters 5 and 6, through the presentation of many examples, we illustrated how some properties and behaviors of their liquids (neat bulk and solution samples) and solids (amorphous and semi-crystalline samples) can be related to the local conformational preferences of their individual chains. In this final chapter, we want, as far as possible, to extend this same ***Inside*** polymer chain microstructure ←→ ***Outside*** polymer material property or function approach to naturally occurring polymers, i.e., to those found in Nature and here called biopolymers.

Our motivation should be obvious. Life as we know and experience is critically dependent on biopolymers. We also wish to add support for the notion that non-polymeric material could not likely be organized into sustainable living life forms. It is difficult to envision the organization of small rigid molecules into living organisms. They lack the ability of single protein chains to change conformations and fold into extended conformations, which can then be further organized into keratin proto-fibrils and eventually fibrils, for example, or fold into globular conformations to allow single protein molecules to act as enzymes or as molecular carriers.

POLYSACCHARIDES

Polysaccharides, such as the starch polymers and cellulose illustrated in Figure 7.1, are the most abundant biopolymers found in Nature. Among these, cellulose is the most abundant organic compound on earth. Think about this the next time you hike through a dense forest or drive by what appears to be an almost endless field of wheat, corn, soybeans, cotton, or some other farm crop. Cellulose serves as a structural element in the cell walls of plants like wood and cotton. Amylose starch, on the other hand, serves as the principal

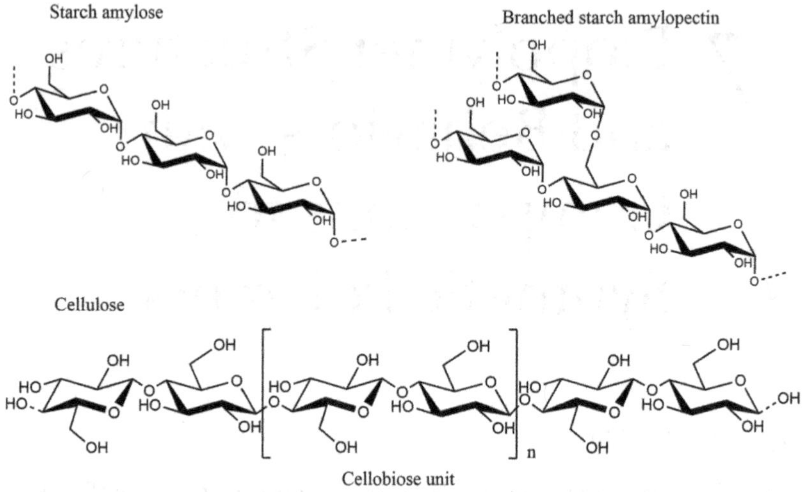

Starch amylose

Branched starch amylopectin

Cellulose

Cellobiose unit

FIGURE 7.1 Linear starch amylose (upper left), branched starch amylopectin (upper right), and cellulose (lower).

food reserve in plants and is the major source of carbohydrates in the diets of animals, including humans, and is generally obtained from wheat, rice, corn, and potatoes. Humans possess amylase enzymes which can randomly cut the backbone bonds in the linear component of starch (amylose) and convert it to sugar (D-glucose) with its associated energy in the form of calories. No such cellulase enzymes exist in humans for digesting cellulose, but are present in other organisms, such as termites, which can literally "eat you out of house and home" if yours is constructed from wood.

The linear starches amylose and cellulose appear to have closely similar structures, i.e., D-glucose rings connected by 1,4 (C_1–O and O–C_4) glycosidic bonds. However, the 1,4 linkages in amylose are all of the α-form, while only 1,4-β-linked D-glucose rings occur in cellulose (see Figure 7.1). On the surface, this appears to be a minor structural difference. However, when it comes to the properties and behaviors of these two linear polysaccharides, which are widely different, it obviously has major consequences. Native cellulose is usually about 70% crystalline, virtually insoluble, and ignites above 280°C before melting (Daniel 1990). On the other hand, amylose starch, though also somewhat crystalline and ignitable before melting, shows some limited solubility in water.

Though the essentially rigid D-glucose sugar rings are 1,4-linked in both of these linear polysaccharides, the conformational flexibility of the 1,4-α-linkage in amylose is much greater than the 1,4-β-linkage in cellulose. For example, the experimentally observed characteristic ratios of their chain dimensions are $C_\infty = \langle r^2 \rangle / nL^2 = 5$ and 36, respectively, for amylose starch and cellulose, where L is the virtual bond spanning the D-glucose ring and connecting glycosidic oxygens [In the case of the virtually insoluble cellulose, the soluble ester and ether derivatives were made, and their dimensions were actually measured] (Sarko 1976; Okano and Sarko 1984).

Brant and Christ were able to reproduce these chain dimensions using the calculated conformational energies $E(\varphi,\psi)$ for rotations about the C_1–O and O–C_4 glycosidic bonds in amylose and cellulose chains (see Figure 7.2 and Table 7.1, Brant and Christ 1990; Brant 1997).

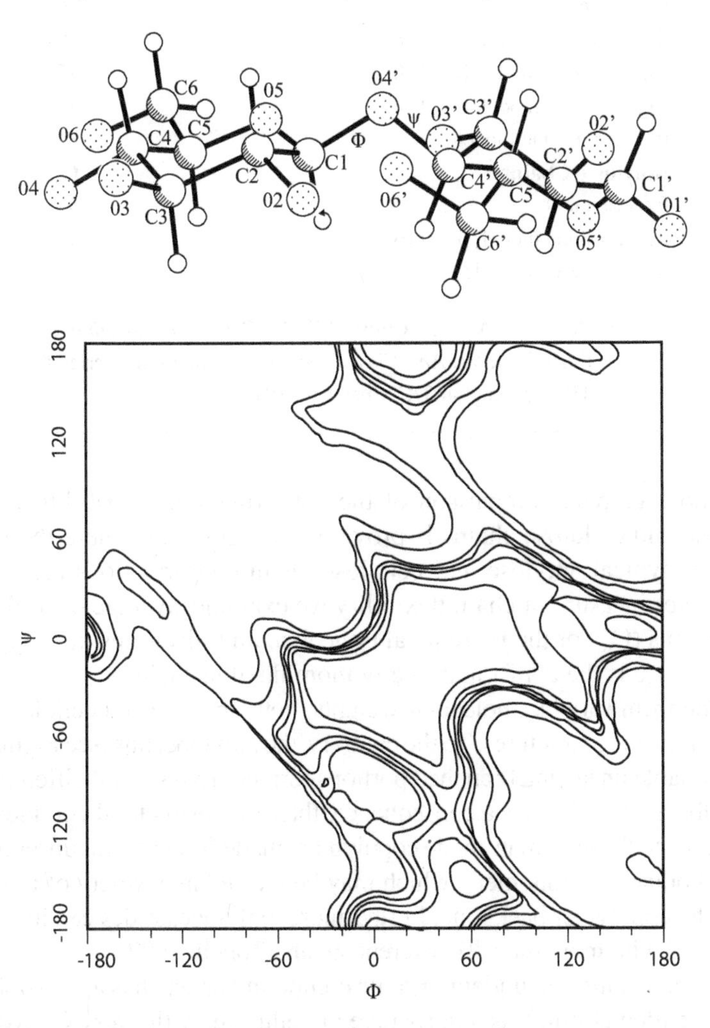

FIGURE 7.2 Upper—Ball and stick drawing of O-4-(β-D-glucopyranosyl)-β-D-glucose ("β-cellobiose"). Glycosidic torsion angles φ and ψ = 0°. Lower—Cellobiose conformational energy map, $E(\varphi,\psi)$, rigid glucose β = 124°, with contours drawn at 2, 4, 6, 8, 10, 25, and 50 kcal/mol above the absolute minimum located near φ,ψ = 10°,–40°. (Reproduced with permission from Brant, D. A., and Christ, M. D., *Computer Modeling of Carbohydrates*, *Amer. Chem. Soc.*, ACS Symposium Series # 430, Washington, DC, Chapter 4, 1990.)

TABLE 7.1
Comparison of Amylose and Cellulose Chain Dimensions and Their Dependence on Chain Geometry (<C_1-O-C_4) and Sugar Ring Flexibility

Cellulosic Chains	C_∞
Experimental	36
Rigid residue model, β = 116°	107
Rigid residue model, β = 120°	75
Rigid residue model, β = 124°	32
Relaxed residue model	11
Amylosic Chains	C_∞
Experimental	5
Rigid residue model, β = 116°	4.5
Relaxed residue model	3

Source: Brant, D. A., and Christ, M. D., *Computer Modeling of Carbohydrates, Am. Chem. Soc.*, ACS Symposium Series # 430, Washington, DC, Chapter 4, 1990.

A more graphic comparison of the conformational flexibilities of the amylose and cellulose chains is provided in Figure 7.3, where the trajectories of typical amylose and cellulose chain conformations are shown. Whichever measure of chain flexibility we examine, average overall chain dimensions (C_∞) or the more local typical chain trajectories, it is apparent that amylose is the conformationally more flexible chain.

Aside from variable molecular weights, both amylose and cellulose have regular repeating structures. To the contrary, the amylopectins (see Figure 7.1) have variable linear and branched portions formed from several different sugars (Alba et al. 2018). As a consequence, their microstructural complexity is analogous to those of man-made copolymers made from several different linear and branched monomers, which may be located in a variety of positions along the copolymer backbone. Furthermore, polysaccharides can in general be formed with more than 30 different sugars (Tonelli 1989).

The second most abundant organic compound on earth is also a polysaccharide called chitin. It is a derivative of cellulose, with every C_2 hydroxyl group replaced by an *N*-acetyl group

$$—\overset{H}{N}—\overset{O}{\overset{\|}{C}}—CH_3.$$

It is primarily found in mushroom cell walls and in the protective outer layer of crustacean and insect shells. Like cellulose, chitin is intractable

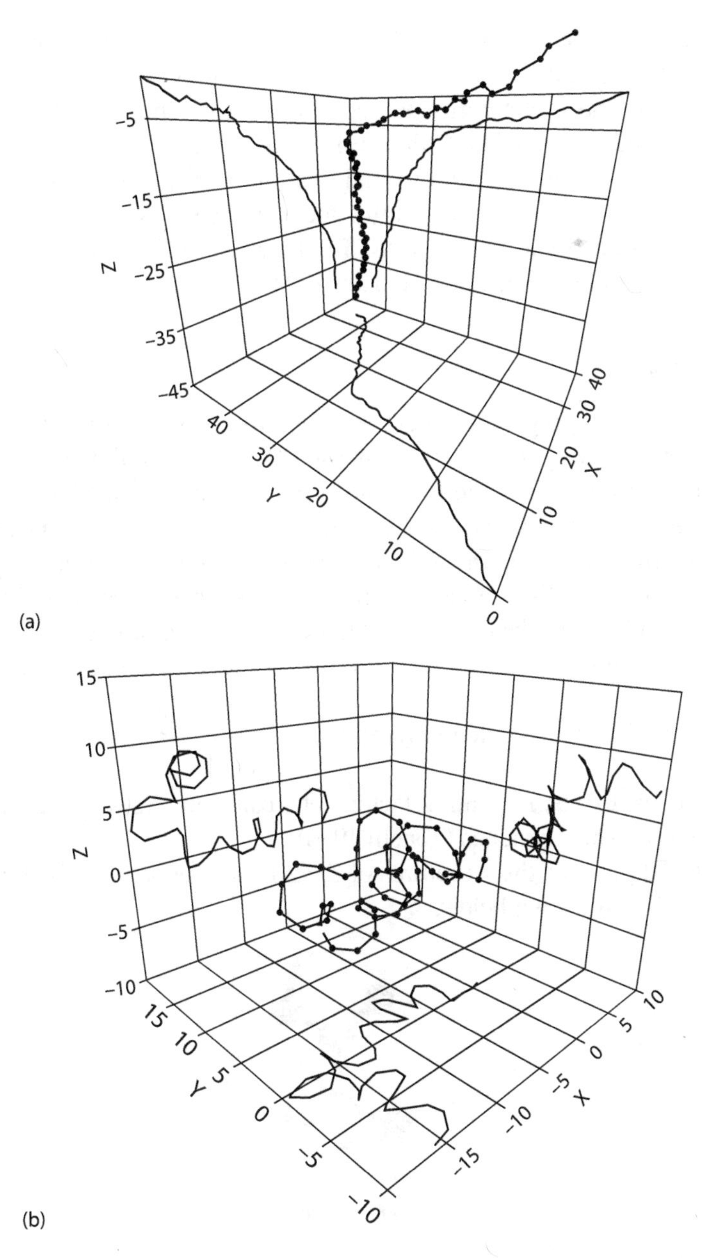

FIGURE 7.3 Typical conformations of cellulose (a) and amylose (b) chains obtained from Monte Carlo sampling of their conformational energy maps. Glycosidic oxygens are represented by •, and they are linked to virtual bonds (not shown) that span neighboring sugar rings. Chain projections onto mutually orthogonal planes aid in the visualization of chain trajectories. (Reproduced with permission from Brant, D. A., and Christ, M. D., *Computer Modeling of Carbohydrates*, *Amer. Chem. Soc.*, ACS Symposium Series # 430, Washington, DC, Chapter 4, 1990.)

and does not melt or dissolve. However, it can be readily converted to chitosan by treatment with alkali (NaOH), which converts the N-acetyl groups to amine ($-NH_2$) groups and confers solubility in solvents such as formic and acetic acids. Chitosan is biodegradable and bioabsorbable, accelerates blood clotting and wound healing, germinates seeds, and can remove heavy metals from contaminated process streams, so it is finding increasing commercial applications (Tokura et al. 1987; Hirano et al. 1989; Rorrer et al. 1993; Rathke and Hudson 1994).

PROTEINS

Proteins are the polymers Nature employs to bring life to animals and plants. They are responsible for the physical structures of living organisms, as well as for all of their life functions. Proteins may be called and considered as nylon-2 copolymers, because their backbones consist of amide bonds separated by a single tetrahedral sp^3 carbon to which is also bonded a proton and 19 different substituent groups R, except for the glycine amino acid residue with R = H. They are produced by the reaction of L-α-amino acids (see below), and, again except for glycine, are chiral molecules. However, certain small proteins called polypeptides, $H_2N{-}CH(R){-}C(=O){-}OH$, that function as hormones, toxins, regulating factors, etc. can sometimes contain one or more D-amino acid residues (Tonelli 1986).

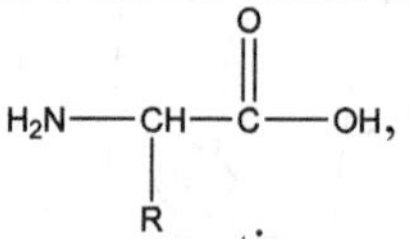

Table 7.2 presents the structures of the 20 α-amino acids, which are all L in chirality as shown below.

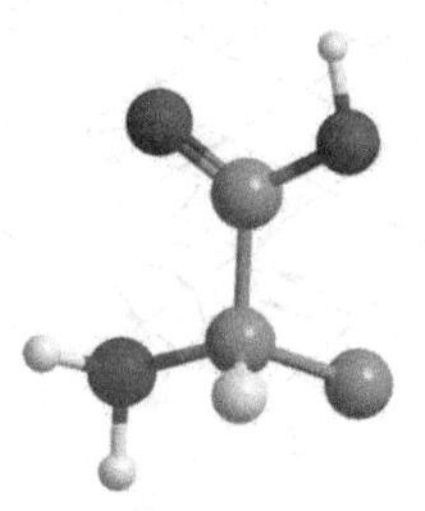

L-Amino Acid

Proteins provide the structural materials holding living organisms together, catalyze virtually all metabolic processes, create antibodies to combat bacterial and viral infections, and transport molecular and ionic species. The versatility of their functions is a consequence of the almost infinite diversity of their potential chemical microstructures. As proteins are built up from the 20 different α-amino acid building blocks, at each position (1,2,3,….n−2, n−1, n) along the protein chain any of the 20 amino

TABLE 7.2

α-Amino Acids Used to Form Proteins*

A—Alanine (Ala)	$H_2N-CH(CH_3)-C(=O)-OH$
R—Arginine (Arg)	$H_2N-CH(CH_2-CH_2-CH_2-NH-C(=NH)-NH_2)-C(=O)-OH$
N—Asparagine (Asn)	$H_2N-CH(CH_2-C(=O)-NH_2)-C(=O)-OH$
D—Aspartic Acid (Asp)	$H_2N-CH(CH_2-C(=O)-OH)-C(=O)-OH$
C—Cysteine (Cys)	$H_2N-CH(CH_2-SH)-C(=O)-OH$
E—Glutamic Acid (Glu)	$H_2N-CH(CH_2-CH_2-C(=O)-OH)-C(=O)-OH$
Q—Glutamine (Gln)	$H_2N-CH(CH_2-CH_2-C(=O)-NH_2)-C(=O)-OH$
G—Glycine (Gly)	$H_2N-CH(H)-C(=O)-OH$
H—Histidine (His)	$H_2N-CH(CH_2$-imidazole ring (N, NH)$)-C(=O)-OH$
I—Isoleucine (Ile)	$H_2N-CH(CH(CH_3)-CH_2-CH_3)-C(=O)-OH$

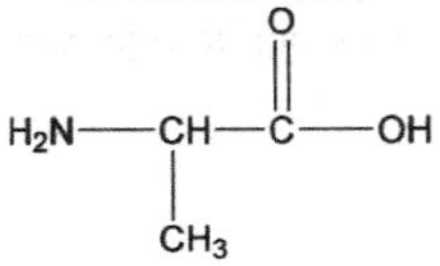

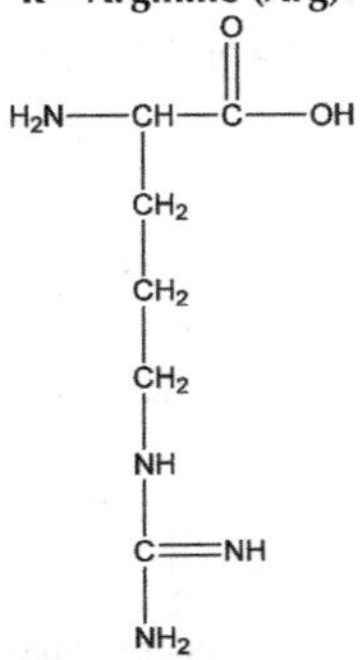

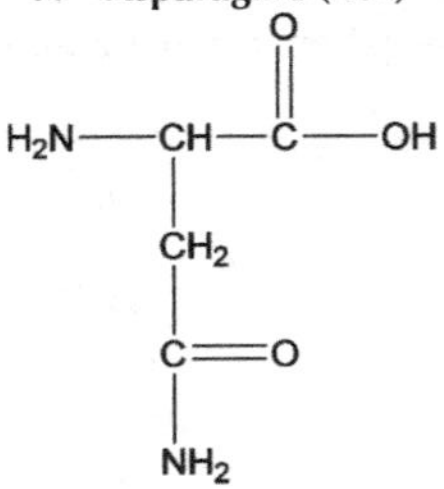

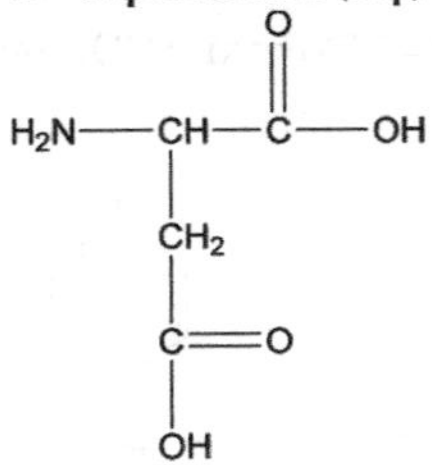

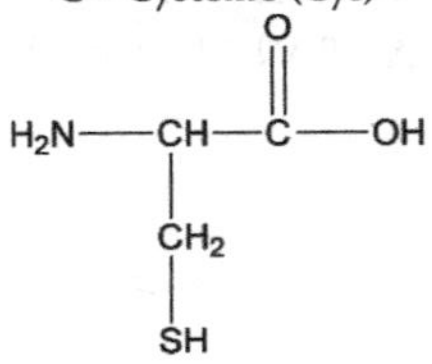

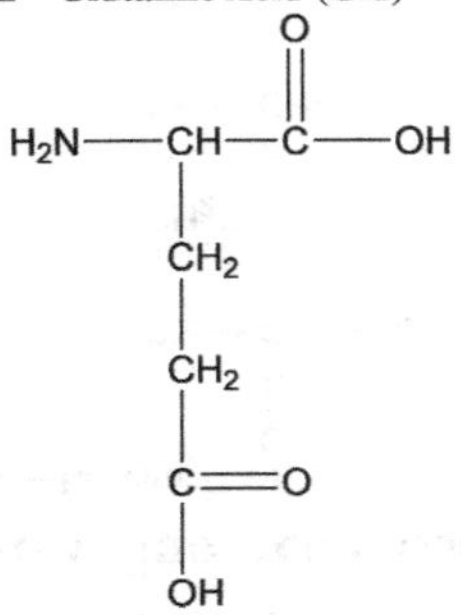

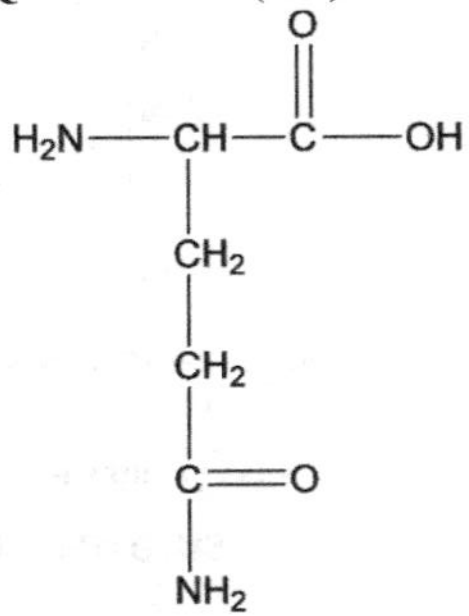

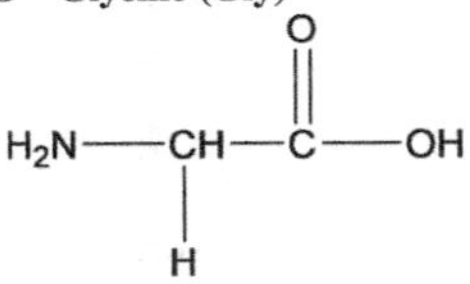

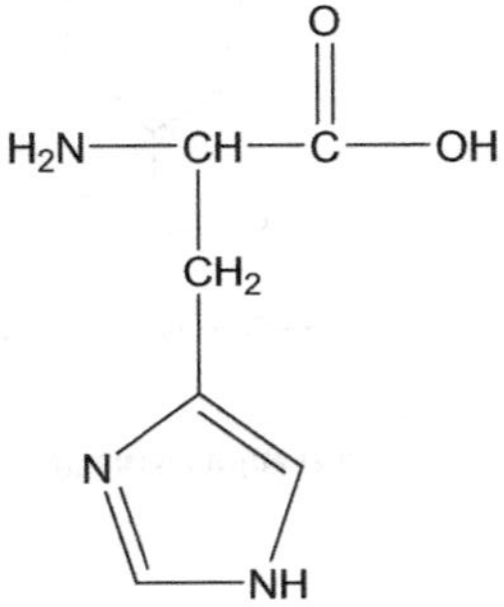

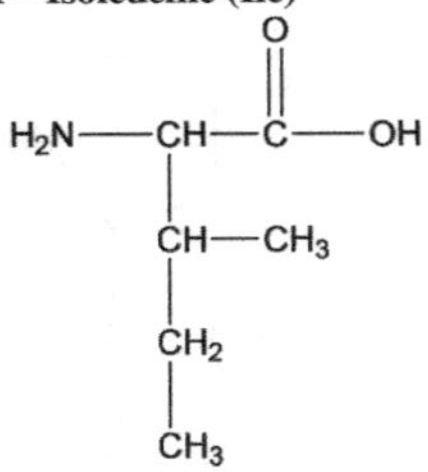

(Continued)

TABLE 7.2 ***(Continued)***

α-Amino Acids Used to Form Proteins*

L—Leucine (Leu)	**K—Lysine (Lys)**	**M—Methionine (Met)**	**F—Phenylalanine (Phe)**	**P—Proline (Pro)**
S—Serine (Ser)	**T—Threonine (Thr)**	**W—Tryptophan (Trp)**	**Y—Tyrosine (Tyr)**	**V—Valine (Val)**

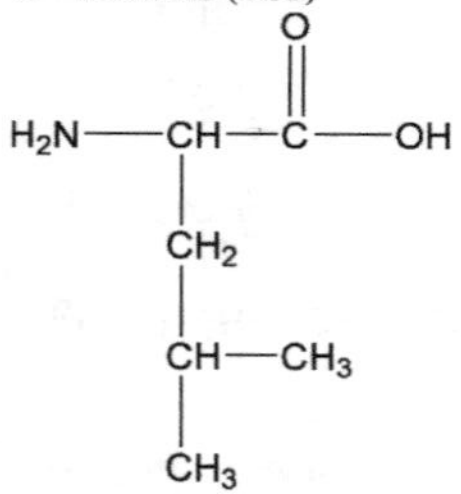

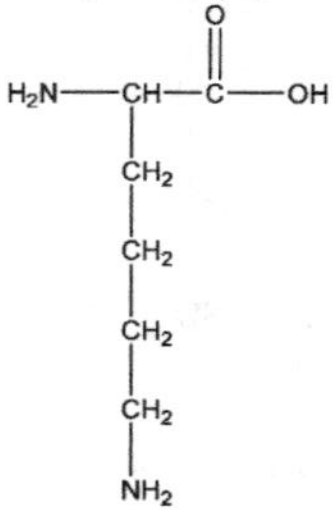

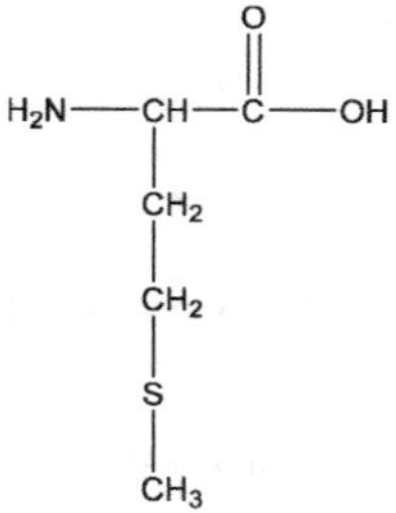

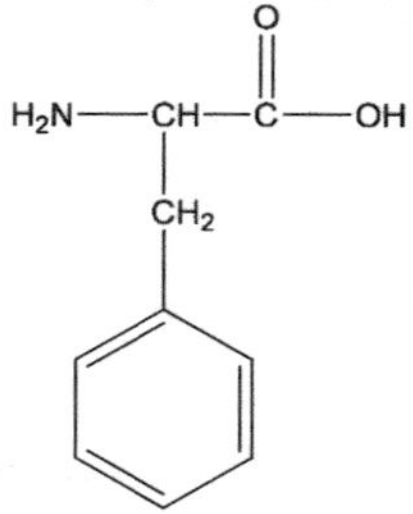

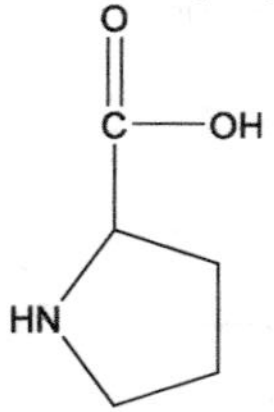

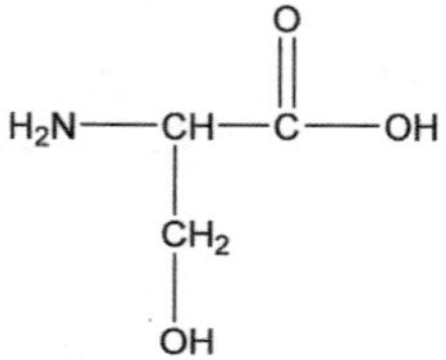

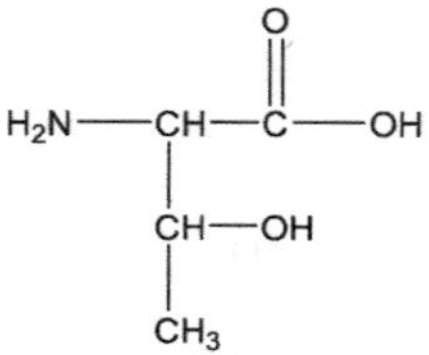

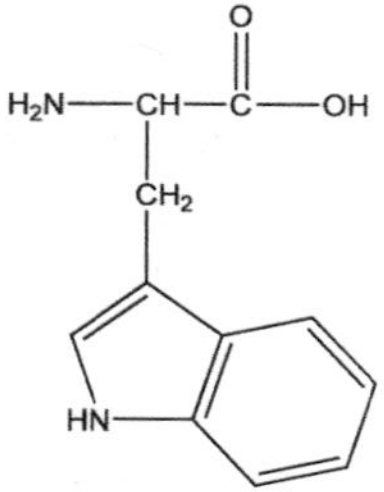

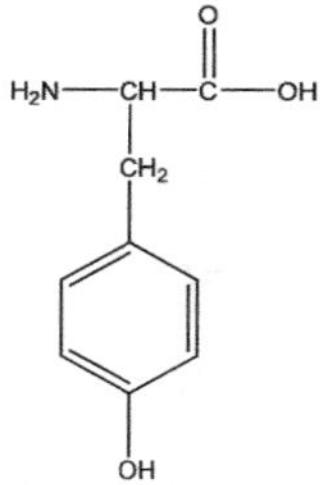

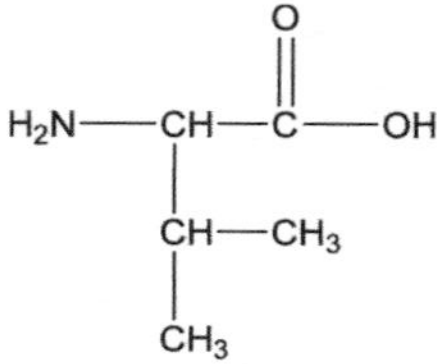

* We have chosen the 3-letter code for the amino acids rather than the single letter code [**G**—Glycine (Gly), **P**—Proline (Pro), **A**—Alanine (Ala), **V**—Valine (Val), **L**—Leucine (Leu), **I**—Isoleucine (Ile), **M**—Methionine (Met), **C**—Cysteine (Cys), **F**—Phenylalanine (Phe), **Y**—Tyrosine (Tyr), **W**—Tryptophan (Trp), **H**—Histidine (His), **K**—Lysine (Lys), **R**—Arginine (Arg), **Q**—Glutamine (Gln), **N**—Asparagine (Asn), **E**—Glutamic Acid (Glu), **D**—Aspartic Acid (Asp), **S**—Serine (Ser), **T**—Threonine (Thr)], because it does not require memorization.

acids may potentially be incorporated. So, for example, proteins consisting of 100 amino acids can have as many as ~10^{130} distinct primary structures, because:

$$\underbrace{20 \times 20 \times 20 \cdots\cdots \times 20 \times 20 \times 20}_{n=100} = 20^{100} \approx 10^{130}$$

(Dickerson and Geis 1969; Bell and Bell 1988). However, Nature selects very specific primary structures to accomplish all of the many life tasks proteins perform.

In addition to their structural diversity, many of the amino acid residues in proteins are able to assume a large variety of conformations or secondary structures, and this also contributes to the extraordinary range of their functions (Flory 1969; Grassberger et al. 1998). Amide bonds possess some double bond character and so are essentially rigid and usually fixed in the trans conformation. The other two single bonds (see below) are, however, in most amino acid residues able to assume a wide range of conformations, which, as in synthetic polyamides and polyesters, are largely independent of the conformations adopted by their neighboring residues (Flory 1969).

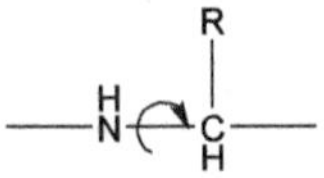

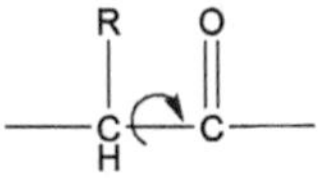

R H N C H R O C H C

In Figure 7.4, a portion of a protein whose conformation depends only upon the conformation of a single amino acid residue is presented, along with some further geometrical structural details. When semi-empirical potential functions similar to those described and used in Chapter 3 for poly(2-vinyl pyridine) are applied to the protein fragment in Figure 7.4, the conformational energy maps for glycine (R = H) and L-alanine (R = CH_3) residues shown in Figure 7.5 are obtained (Brant et al. 1967).

The conformational diversity of the glycine and L-alanine amino acid residues are apparent from their conformational energy maps. It is clear that neither of their rotatable backbone bonds (shown above) are confined exclusively to the staggered *trans, gauche*$^+$, or *gauche*$^-$ conformations (see Figure 3, in Chapter 3). This is a consequence of their low (~1 kcal/mole) inherent rotational barriers (Brant and Flory 1965), which may be contrasted to the (~3.5 kcal/mole) barrier inherent to rotation about sp^3–sp^3 -C—C- bonds (see below) in vinyl polymers (Mizushima 1954; Wilson 1959; Herschback 1963; Bonham and Bartell 1959; Kuchitsu 1959; Bartell and Kohl 1963).

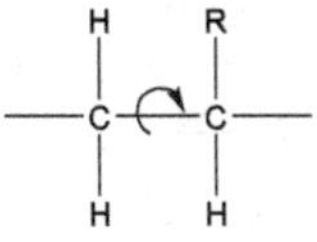

H R C C H H

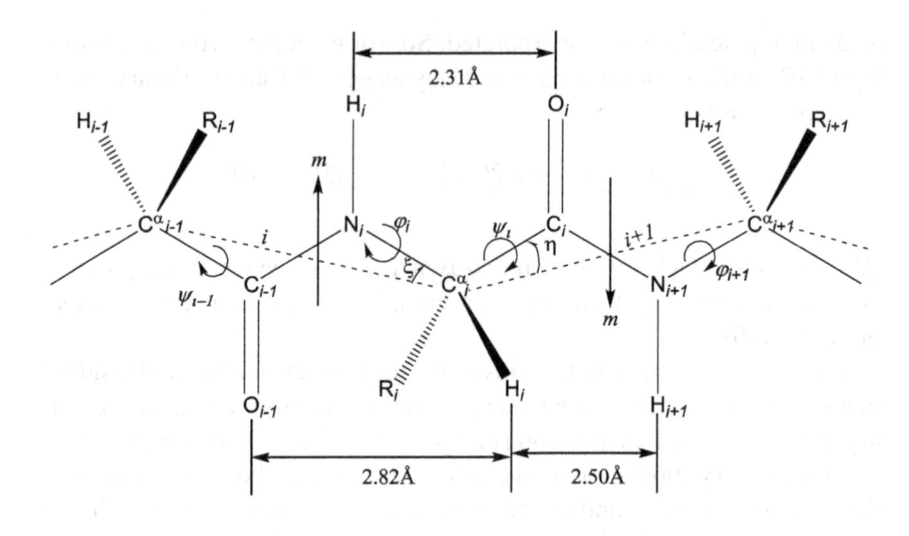

Bond lengths, Å		Bond angle and its supplement (θ), deg	
C^{α}—C	1.53	∠C^{α}CN	66
C—N	1.32	∠CNC^{α}	57
N—C^{α}	1.47	∠NC^{α}C	70
C=O	1.24		
N—H	1.00		
C^{α}—C^{β}	1.54		
C^{α}—H^{α}	1.07		

η=22.2°; ξ=13.2°; l_u=length of virtual bond=3.80Å

C^{β} denotes the first carbon of the substituent R, assuming R to be —CH_2—R′

FIGURE 7.4 Three-residue protein chain whose C^{α}_{i-1} <-> $C^{\alpha}_{i}+_1$ conformation depends only upon the conformation of the central residue shown in the planar $\varphi_i = \psi_i = 0°$ conformation. The amide bond dipole moments *m* are shown and the bond lengths and valence angles are listed below the figure. Note the dashed lines between neighboring asymmetric C^{α}s have a constant length l_{μ} = 3.80 Å that is independent of residue conformation (φ_i,ψ_i). Thus, l_{μ} is a virtual bond, and with conformational averaging over their energy maps [$E(\varphi_i,\psi_i)$ vs. (φ_i,ψ_i)], like those shown below in Figure 7.5, the dimensions of randomly coiling homo-polypeptides and proteins (see below) may be estimated. (From Flory, P. J., *Statistical Mechanics of Chain Molecules*, Wiley-Interscience, New York, 1969.)

Similar to synthetic polymers, dissolved proteins and homo-polypeptides may be denatured from their native states and assume randomly coiling conformations, which may be characterized using conformational energy maps of their amino acid residues like those in Figure 7.5 (Brant et al. 1967). This has been shown to lead to characteristic ratios $C_n = \langle r^2 \rangle_o / n l_{\mu}^2$, where n and l_{μ} are, respectively, the number of amino acid residues and the length of the virtual bond connecting the C^{α} carbons

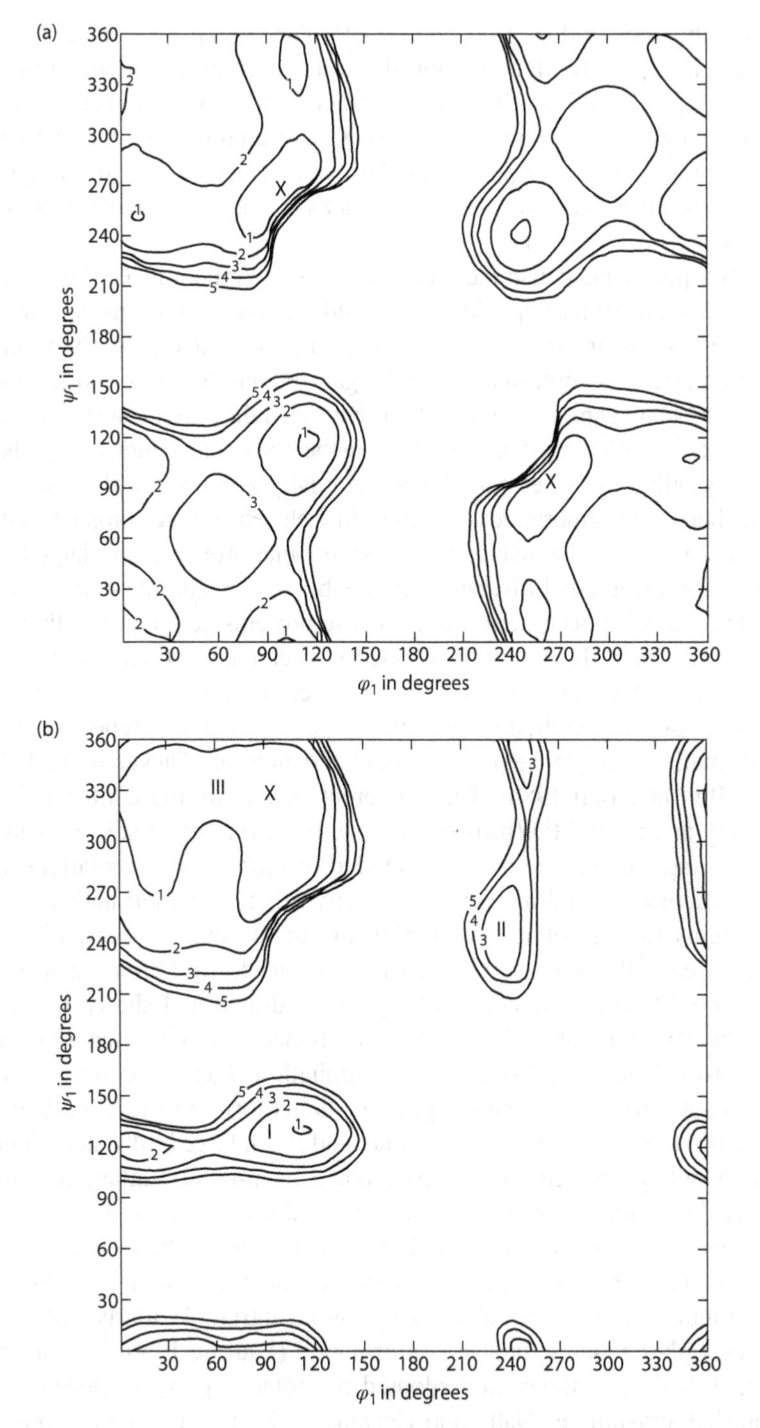

FIGURE 7.5 Conformational energy maps for glycine (a) and L-alanine (b) residues. (Reproduced with permission from Brant, D. A. et al., *J. Mol. Biol.*, 23, 47, 1967.)

in neighboring residues (see Figure 7.4). The dimensions calculated for polyglycine, poly (L-alanine), and denatured silk protein (β-keratin) are $C_n = \langle r^2 \rangle_o / nl_\mu^2 = 2.2$, 9.3, and 2.4 (Brant and Flory 1965; Mathur et al. 1997). Polypeptides with R = $-CH_2X$ side-chains have experimental dimensions of $C_n = 9.0 \pm 0.5$ and the average dimension measured for denatured silk protein was $C_n = 2.4$ (Jackson and O'Brien 1995; Mathur et al. 1997).

Polypeptides and denatured proteins are clearly randomly coiling in solution, with high intimacy quotients (IQs) and average dimensions that are predictable from their conformational energy maps. Native proteins, whether adopting extended conformations and organized into fibers or folded up into dense globular shapes, have IQs of ~1, but when they are denatured assume random coils with large IQs. The chains in the crystalline regions of synthetic semi-crystalline polymer samples are similarly extended with small IQs, while those in the amorphous regions or in melts have more compact conformations, but unlike the much more dense globular proteins, have large IQs.

The conformational diversity and flexibility of randomly coiling denatured proteins is not so important, because proteins do not generally function as random-coils. Rather, they function as elements of fibrous structural constructs with regular repeating extended conformations, or as enzymes, transport agents, antibodies, etc. with specific, but not repeating local conformations and dense overall globular structures (Dickerson and Geis 1969; Bell and Bell 1988). Their inherent structural and conformational diversity and flexibilities provide them with the important ability to adopt a wide range of specific overall extended or globular conformations (tertiary structures), enabling them to uniquely function in their native states.

Proteins that adopt regular highly extended conformations like silk fibroin generally interact with each other to form fibrous structures. In Figure 7.6, fully extended polyglycine is drawn and shown to represent the structure of hydrogen-bonded β-sheet types of materials like silk fibroin. The crystal structures of aliphatic nylons are closely similar (see Chapter 6). Notice that the protein chains forming hydrogen-bonded β-sheets cannot be formed from amino acids with large bulky side-chains, which would push them too far apart to hydrogen-bond. Rather, they must be largely composed of glycine, alanine, and serine amino acid residues that have small side-chains (see Table 7.1) to permit their chains to get close enough to hydrogen-bond together into sheets as shown in Figure 7.6.

In Figure 7.7, a protein chain adopting an intramolecularly hydrogen-bonded right-handed α-helical conformation (Pauling 1960) is compared to the 3_1-helical conformation adopted by isotactic-polypropylene chains when they crystallize (Natta and Corradini 1960). The α-helical protein chain is held together and stabilized by intrachain hydrogen bonds between N–H and O=C groups separated by three intervening amino acid residues.

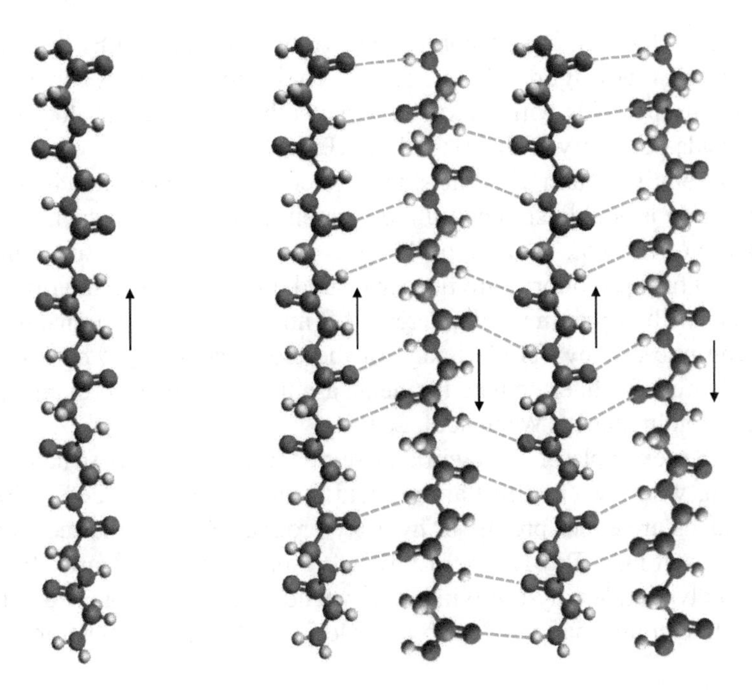

FIGURE 7.6 All *trans* conformation of anti-parallel silk fibroin protein chains hydrogen-bonded to each other into β-sheets (Marsh, R. E. et al., *Biochem. Biophys. Acta*, 16, 13, 1955.) (shown here are polyglycine chains for simplicity).

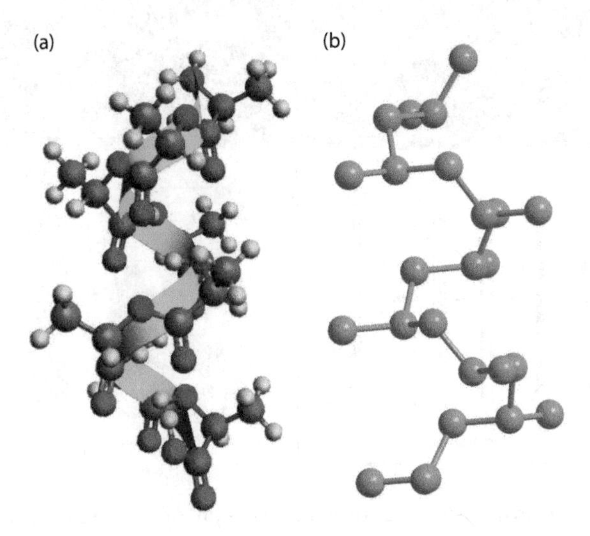

FIGURE 7.7 (a) α-helix polypeptide conformation (Pauling, L., *The Nature of the Chemical Bond*, 3rd Ed., Cornell University Press, Ithaca, New York, 1960.) (shown here is polyalanine chain for simplicity) and (b) 3_1-helical conformation of crystalline isotactic polypropylene (Natta, G., and Corradini, P. (1960), *Nuova Cimento. Suppl.*, 15, 68–82, 1960.) (hydrogens omitted for clarity).

Also note, that unlike silk fibroin with fully extended chains hydrogen-bonded together (Figure 7.6), the protein chains in α-keratins can accommodate amino acids with bulky side-chains, because they are located on the outside periphery of their α-helices (Figure 7.7).

In α-keratin protein constructs, two α-helical keratin protein chains are wound into a double-stranded dimer protein rope, two of which in turn form proto-filament. Two proto-filaments are then organized into a proto-fibril and four protofibrils are further wound together into an intermediate filament with approximate diameters of 7 nm as seen in the transmission electron microscopy (TEM) image of the cross-section of a human hair (see Figure 7.8). Intermediate filaments are finally embedded in an amorphous protein matrix (Wang et al. 2016).

Another example of a protein that serves to build fibrous structures is collagen, whose every third amino acid residue is glycine, 20%–25% of its residues are either proline or hydroxy-proline (HyPro), and most of its sequences are Gly-Pro-HyPro or Gly-X-Pro or HyPro. In Figure 7.9a, the homo-polypeptide poly-L-proline is shown adopting its left-handed helical crystalline conformation (Cowen and McGavin 1955; Sasisekharan 1959;

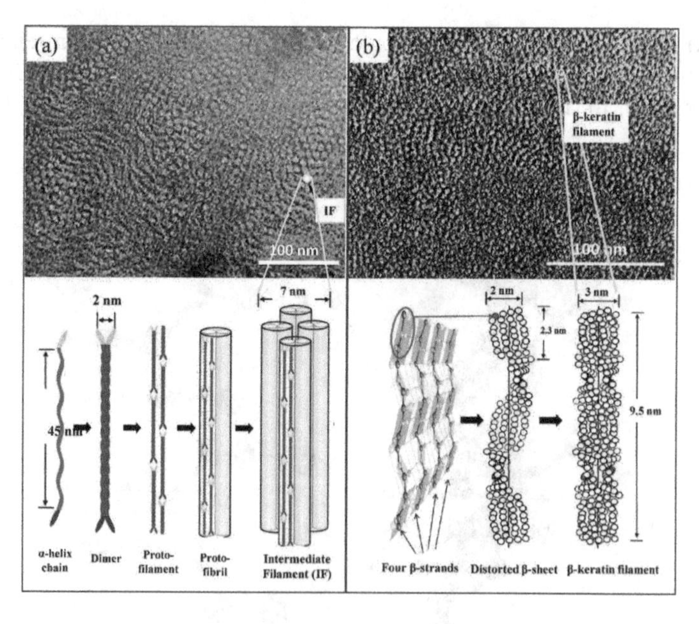

FIGURE 7.8 Transmission electron micrograph of the cross-section of (a) a human hair (α-keratin) and (b) seagull feather rachis (β-keratin), and their schematic structural details. (Adapted with permission from Wang et al., 2016; Fraser, R. D. B. et al., *Keratins: Their Composition, Structure and Biosynthesis*, Charles C. Thomas, 1972; Lodish, H. et al., New York. *Molecular Cell Biology*, Macmillan, 2008; Voet, D. et al., *Fundamentals of Biochemistry: Life at The Molecular Level*, John Wiley & Sons, 2008.)

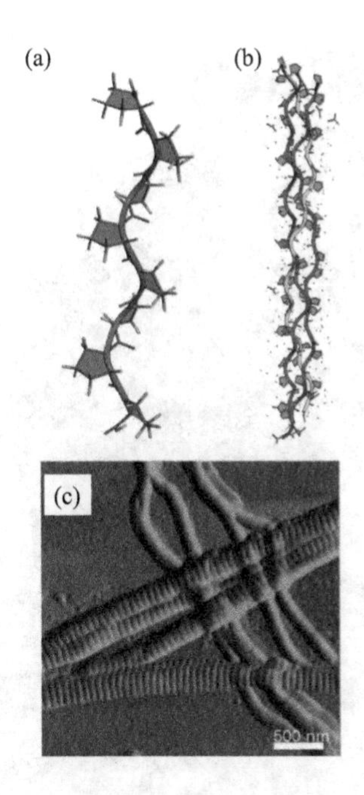

FIGURE 7.9 (a) The homo-polypeptide poly-L-proline is shown adopting a left-handed helical conformation adopted when it crystallizes. (Cowen and McGavin 1955; Sasisekharan 1959; Sarko 1976). (b) Three-stranded "tropocollagen" fiber. (Image created using Avagadro software, and X-ray diffraction data with PDB ID 1CGD). (From Bella, J. et al., Structure, 3, 893–906, 1995). (c) Atomic force microscopy image of calfskin collagen fibrils. (Adapted with permission from Li, Y., and Douglas, E. P., *Colloids Surf. B.*: Biointer., 112, 42–50, 2013.)

Sarko 1976), which can interact, intertwine, and hydrogen-bond with two additional poly-L-proline chains to form a three-stranded "tropocollagen" fiber in Figure 7.9b. Calfskin collagen is shown in Figure 7.9c.

Collagen proteins make up the fibrous components of tendons, cartilage, and ligaments. It also serves as a scaffold upon which the mineral hydroxy-apatite deposits, crystallizes, and forms bone (Matheja and Degens 1971).

We have seen that structural proteins are able to organize and pack into fibrous structures, because they can adopt extended regular repeating conformations. Semi-crystalline synthetic polymers can likewise adopt regular repeating extended low energy conformations and crystallize. They may also be spun into fibers and evidence physical properties similar to those of the structural proteins (Mandelkern 1972; Tonelli 1970, 1974).

Globular proteins, like lysozyme shown in Figure 7.10, adopt folded compact conformations and overall sizes and shapes not assumed by synthetic polymers (Flory 1969; Mandelkern 1972). Tightly folded compact conformations with very little unoccupied volume within their spatial

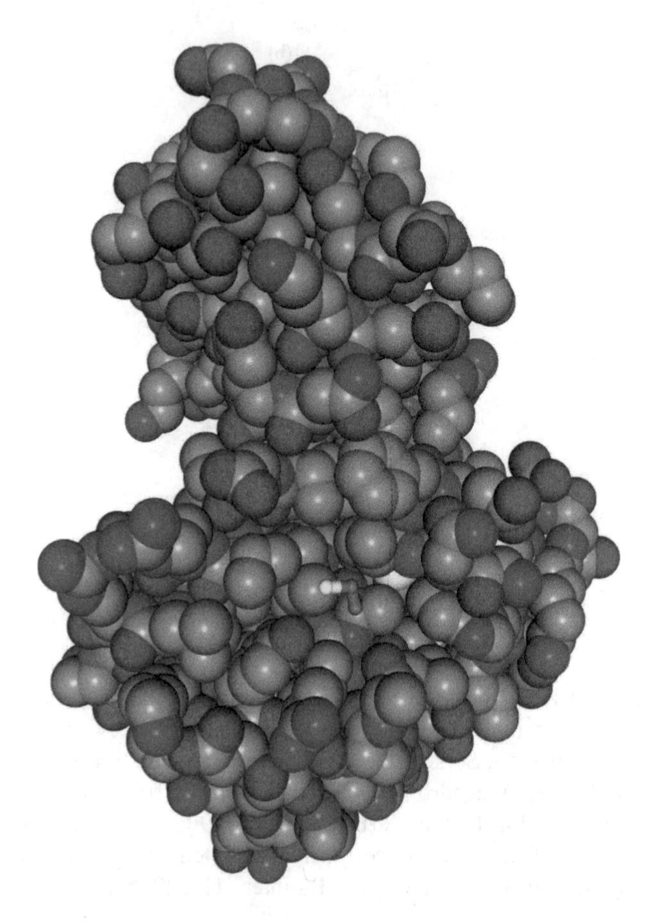

FIGURE 7.10 A space-filling molecular model of the globular protein lysozyme, an anti-microbial animal enzyme forming part of the immune system that catalyzes the hydrolysis of gram-positive bacterial cell walls. Image created using NGL Viewer (Rose et al. 2018, NGL viewer: web-based molecular graphics for large complexes. Bioinformatics doi:10.1093/bioinformatics/bty419), and X-ray diffraction data with PDB ID 253L (From Shoichet, B. K. et al., *Proc. Natl. Acad. Sci. USA* 92, 452–456, 1995.) on RCSB PDB (http://www.rcsb.org/).

domains lead to IQs approaching unity and are the norm for globular proteins. However, as noted above, when denatured, they unfold and adopt randomly coiled conformations with high IQs similar to dissolved or molten synthetic polymers. The external influence of native tightly coiled globular proteins is essentially limited to the volumes they physically occupy, thereby limiting the cooperative nature of their behaviors.

Based on single crystal X-ray diffraction data, a space-filling model of lysozyme was constructed and is presented above in Figure 7.10 (Shoichet et al. 1995). The digestion of cell walls belonging to gram-positive bacteria is catalyzed by lysozyme, and so it plays an integral role in the immune

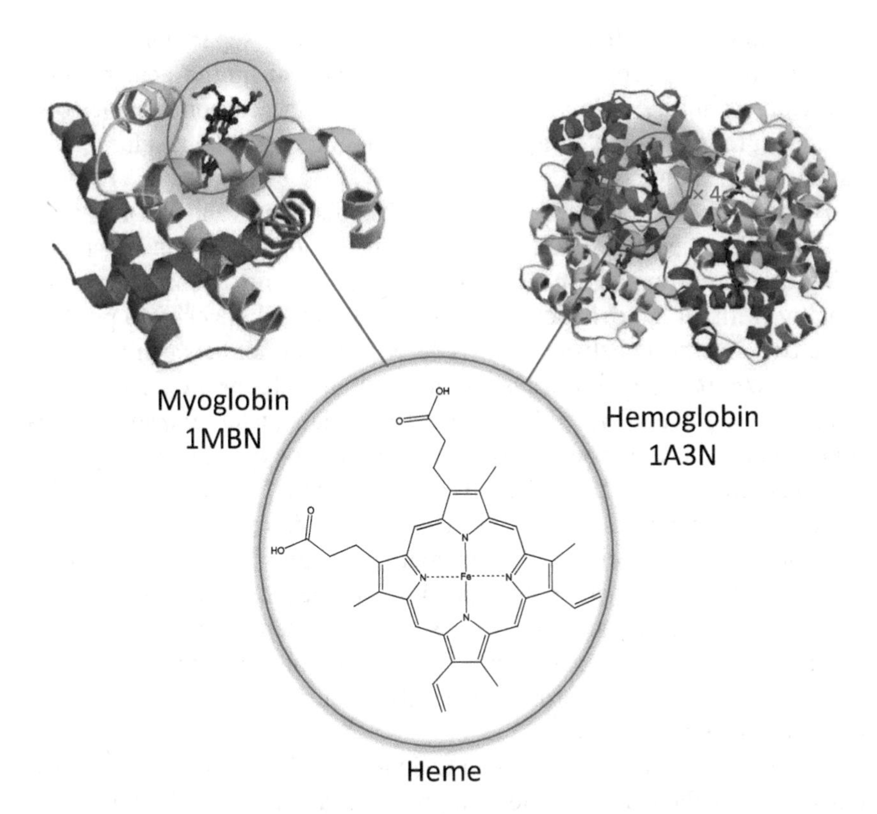

FIGURE 7.11 Schematic representations of the globular structures of myoglobin and hemoglobin and chemical structure of the heme group. 3D protein structure images from the RCSB PDB (www.rcsb.org) of PDB ID 1MBN (Watson 1969) and 1A3N (Tame and Vallone 2000).

defenses of animals. Notice the tight packing of atoms (IQ ~ 1), which coupled with limited protein interactions, often leads to the formation of complete crystallization and the production of single crystals almost never seen with synthetic polymers, which are rather almost always semi-crystalline. Both of these characteristics of globular proteins lead to their nearly identical solution and solid-state native conformations and structures.

In Figure 7.11, schematic representations of the crystalline structures of the transport proteins myoglobin and hemoglobin are presented, together with the structure of their constituent heme groups that are reversibly able to bind oxygen and carbon dioxide. In the lungs, hemoglobin binds oxygen molecules and delivers it to myoglobin in the muscles. There, it is stored and used when needed for oxidative metabolic processes, which produce carbon dioxide. The carbon dioxide is transported by the myoglobin in the muscles to hemoglobin in the bloodstream and is eventually exhaled from the lungs.

Myoglobin's overall globular shape is achieved by separation along the protein chain and appropriate folding of α-helical amino acid regions. Myoglobin's ability to bind and release oxygen and carbon dioxide from its heme group is achieved by the proper folding of its 153 residues into a carrier for the heme group.

Hemoglobin is made from two pairs each of distinct α- and β-polypeptide chains with, respectively, 141 and 146 amino acid residues and each carrying a heme group. Each α, β subunit resembles myoglobin conformationally, with a high α-helical content, and are closely packed to produce close $\alpha \cdots \beta$ contacts between their side-chains, but not between the same subunits $\alpha \cdots \alpha$ or $\beta \cdots \beta$. Because α and β subunit chains are packed symmetrically, an overall roughly spherical globular quaternary structure is produced and permits control over the amount of oxygen the heme groups bind and transport. The amazingly sophisticated functions of myoglobin and hemoglobin directly result from their microstructures, i.e., their sequences of amino acids or their primary structures, as do the properties and behaviors of synthetic polymers. However, because proteins are made from 20 distinct amino acids with specific sequences, while synthetic polymer structures are much simpler and more homogeneous, the range of behaviors/functions of proteins is vastly greater and much more specific and sophisticated.

It is the sequences of amino acid residues or primary structures of proteins and not their amino acid compositions alone that determine their amazingly diverse biological functions. To a more minor extent, synthetic polymers share this trait (think regularly alternating vs. random comonomer sequences in Chapter 6), but with properties and behaviors that are correspondingly much less specific, elaborate, and varied. The specific connectivity or primary sequence of amino acids in proteins exquisitely controls their biological functions, as illustrated in the following examples.

The first example we offer is the oxygen transporting protein hemoglobin. Hemoglobin proteins in people suffering from sickle-cell anemia have a single mistake in their primary sequence (Watson et al. 1987). Instead of glutamic acid in the 6th position of its β-chains, another amino acid is enchained there, most often valine (see Table 7.1). How can such an apparently minor amino acid substitution $\{2/[(2 \times 141 + 2 \times 146)] \times 100 = 0.34\%$ substitution$\}$ cause such a serious alteration in hemoglobin behavior? It is likely that the substitution of Val for Glu in the vicinity of the 6th residue changes the local conformation of the β-chains, possibly disrupting an α-helical region or the arrangement of neighboring α-helices (protein secondary structure). This in turn would affect the overall folding of the β-chains (protein tertiary structure) and also modify the packing of β- and α-chains (protein quaternary structure), leading to an altered arrangement of heme groups with attendant poorer oxygen and carbon dioxide binding capacities.

Our second example of the sensitivity of protein functions to their primary structures is afforded by Cetacean mammals, including whales, dolphins, and porpoises that have adapted to the marine environment. They have evolved into the Mysticeti (baleen whales) and the Odontoceti (toothed whales) (Zhao et al. 2017). The Cetaceans are also distinguished by a great range in their sizes, anywhere from 1 to 30 meters in length. For example, Cetacea include the blue baleen whale, the largest living mammal with lengths and weights reaching a maximum of approximately 33 m and 190,000 kg, but also the toothed vaquita at 1.4 m in length and a weight less than 40 kg.

Zhao et al. have postulated that an important regulator of metabolism, food intake, and fat storage, by the melanocortin system, plays a pivotal role in Cetacean feeding behaviors and sizes. It had been suggested that dietary specialization may have played a significant role in the evolutionary history of Cetaceans and caused the large variations in their body sizes (Slater et al. 2010). The unique collection of neural circuits in the central melanocortin system is capable of sensing signals from a wide array of hormones, nutrients, and other neural inputs and is primarily received by the hypothalamus. The neural signals are mainly received by the brainstem and are involved in the regulation of hunger and satiety (Chaney 2005 and Cone 2006).

Zhao et al. sequenced the melanocortin-4 receptors of 20 Cetacean species and found a single amino acid difference at the 156th position: Glu for baleen whales and Arg for toothed whales and dolphins. Because the arginine variant of the melanocortin-4 receptor has a higher affinity than the glutamine variant for binding its native ligand, which suppresses appetite, this may explain why the toothed whales are smaller than filter-feeding baleen whales, which have evolved to reach such gargantuan (truly leviathan) sizes. Zhao et al. admit "We know that there are many genes that contribute to size, but we found a single amino acid mutation that fits with the hypothesis that this gene could be one of them."

These two examples make clear that the primary structures of proteins serve to determine their complete structural hierarchy: from their *secondary* structures (local conformations) to their *tertiary* structures (overall folded shape) to their *quaternary* structures (arrangement and packing of *tertiary* folded subunits), with the latter two controlling their biological functions (Anfinsen 1972). This was most straightforwardly demonstrated by the refolding of denatured ribonuclease produced by reductive cleavage of its four disulfide bonds into its fully biologically active native globular conformation (see Figure 7.12). This required a single pairing and reaction of the 105 possible pairings of the eight CySH sulfhydryl groups to form the four native disulfide linkages schematically indicated below in Figure 7.12 (Sela et al. 1957; Anfinsen 1972).

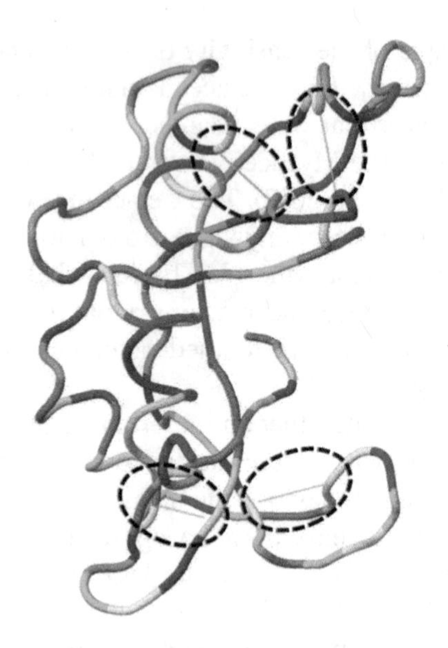

FIGURE 7.12 The chain backbone contour of bovine pancreatic ribonuclease. Disulfide bonds are shown in yellow. PDB ID 1EIC (Chatani et al. 2002), image created by Jmol (http://www.jmol.org/).

More recently it has become apparent that protein folding is not as simple as that just described for the relatively short protein bovine pancreatic ribonuclease. "In the packed, busy confines of a living cell, hundreds of chaperone proteins vigilantly monitor and control protein folding. From the moment proteins are generated in and then exit the ribosome until their demise by degradation, chaperones act like helicopter parents, jumping in at the first signs of bad behavior to nip misfolding in the bud or to sequester problematically folded proteins before their aggregation causes disease" (Wilson and Clark 2016; Everts 2017). It is also believed that the redundant nature of protein synthesis may also play a role in protein folding.

As we shall soon discuss, only the proteins necessary for a particular living organism are synthesized in a specific templated manner directed by the information carried by an organism's DNA. Polynucleotides consist of a sugar-phosphate backbone with one of four distinct bases (A = adenine, T = thymine, G = guanine, C = cytosine in DNA) attached to their sugars (Watson et al. 1987; Frank 1999). A sequence of three consecutive nucleotide units, with an overall $4 \times 4 \times 4 = 64$ possible distinct base sequence triplets comprise the genetic code or codons corresponding to each of the 20 amino acids, as well as codons for beginning and halting the synthesis of each distinct protein required by the organism. This raises the question of what is the purpose of the ~42 (64–20–2) extra codons? We know that many of the amino acids are coded by more than a single three base codon, e.g., phenylalanine and serine have, respectively, two (UUU and UUC) and four (UCU, UCC, UCA, UCG) codons.

Recent observations are suggesting that some redundant codons result in fast protein synthesis, while others serve to slow it down. The "slow" codons often occur between polynucleotide regions that code for protein segments that need to fold independently (Everts 2017). The pauses during protein synthesis that these "slow" codons may enable "is like a stutter that allows individual regions (of a protein) to fold" independently (Marqusee in Everts 2017; also Chaney et al. 2017).

Despite the very large numbers of globular proteins whose sequences of amino acids and 3-dimensional native structures have been determined, we have not been able to successfully correlate them to confidently predict or connect the tertiary structures of proteins to their primary structures (Grassberger et al. 1998; Everts 2017; Ovchinnikov et al. 2017). Despite the much simpler microstructures of synthetic polymers, determination of their overall molecular structures also remains a challenge. Though as noted in Chapter 4, ^{13}C-NMR in many instances enables the determination of the types and quantities of their constituent short-range microstructures (Tonelli 1989), we still do not know where along the backbones of synthetic polymers each of their microstructures is located (Gurarslan and Tonelli 2017). This is equivalent to knowing the amino acid composition of a protein, but without concomitant knowledge of its primary structure (Babu et al. 2011; Babu 2016).

The relative structural simplicity of synthetic polymers precludes them from individually folding into compact globules, but they can adopt extended repeating conformations and interact with each other to form homopolymer crystals or stereocomplex crystals as observed for syndiotactic and isotactic poly(methyl methacrylate)s (s-,i-PMMAs) or poly L- and D-lactides (PLLA, PDLA). Figure 7.13 shows the suggested 2:1 s-PMMA:i-PMMA and the 1:1 PLLA:PDLA complexes (Kumaki et al. 2007; Tsuji 2016).

Very recently it has been discovered that some proteins neither adopt extended nor globular tertiary structures (Iakoucheva et al. 2002; Uversky et al. 2008; Babu et al. 2011; Gsponer and Babu 2012; Tompa 2012; Wright and Dyson 2015; Babu 2016). Rather, they function without well defined 3-dimensional structures, as intrinsically conformationally disordered proteins. As the first step toward understanding their diverse cellular functions, course-grain molecular modeling has begun to be used to determine the ensemble of conformations that intrinsically disordered proteins explore under physiological conditions (Baul et al. 2019). It is hoped that this information will lead to some understanding of their biological functions.

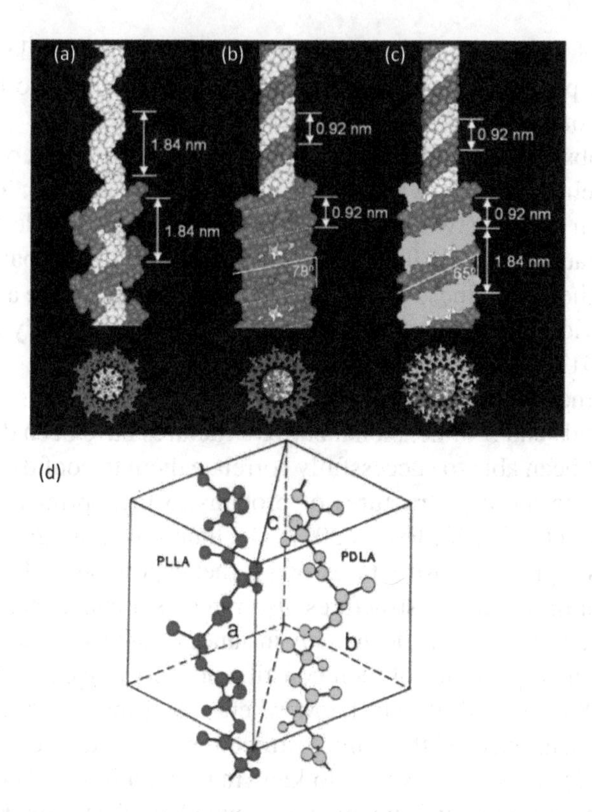

FIGURE 7.13 (a) Double-stranded, (b) triple-stranded, and (c) quadruple-stranded helices proposed for the 2:1 s-PMMA: i-PMMA complex. (From Kumaki, J. et al., *Angew. Chem. Int. Ed.*, 46, 5348, 2007 with permission.) (d) Unit cell of the 1:1 PLLA:PDLA complex crystal. (Taken from Tsuji, H., *Adv. Drug Deliv. Rev.*, 107, 97, 2016 with permission.)

POLYNUCLEOTIDES

The final class of biopolymers we discuss is the polynucleotides, including DNA and RNA that are able to self-replicate and faithfully direct the synthesis of proteins. Their chemical structures are presented in Figure 7.14 and consist of a sugar-phosphate backbone, with phosphate attachment at the 3 and 5 positions on their furanose sugars, deoxy-D-ribose (DNA) and D-ribose (RNA), and several different organic bases attached to the 1-position of their sugar rings. These organic sugars are purines [adenine (A) or guanine (G)] or pyrimidines [cytosine (C) or thymine (T) or uracil (U)]. DNAs differ from RNAs in two structural ways. First, DNAs lack the hydroxyl on the C_2 of the RNA D-ribose sugar rings. Second, the thymine (T) bases on DNAs are replaced by uracil (U) bases on RNAs.

Aside from the hydroxyls on the RNA sugar rings, the backbones of DNA and RNA are identical, so it must be the organic bases attached to their

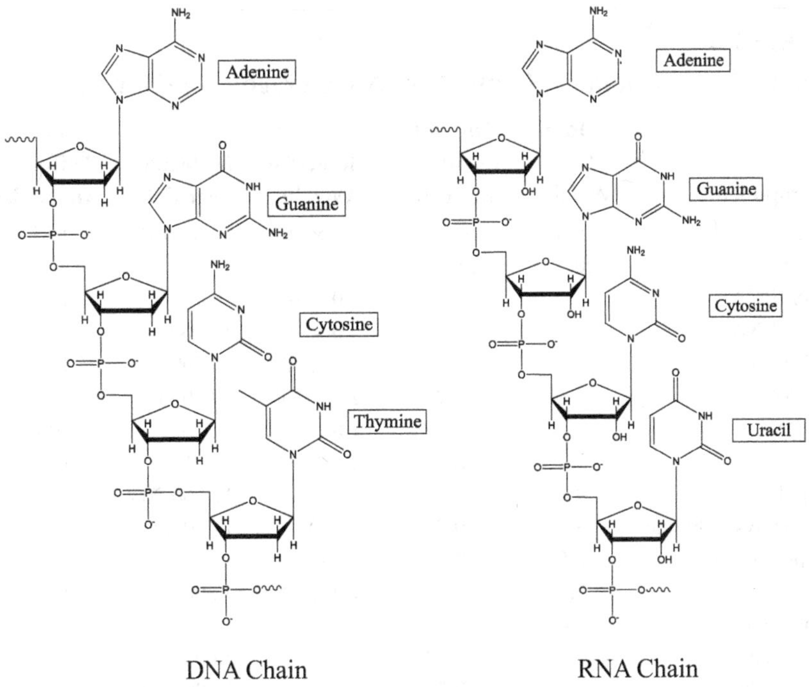

DNA Chain

RNA Chain

FIGURE 7.14 Structures of DNA and RNA polynucleotides.

sugar rings that enable them to self-replicate and direct the syntheses of proteins. Since 20 distinct amino acids must be coded for by these polynucleotides, single bases (4) and neighboring base pairs ($4 \times 4 = 16$) do not contain enough information for coding all of them. The minimum sequence of bases needed for coding all 20 amino acids must be 3, because $4 \times 4 \times 4 = 4^3 = 64$ distinct base triplets that inherently contain more than enough information.

The simple single-cell bacterium *E. coli* needs ~2,000 different proteins to function (see Table 7.3). Assigning an average of 100 to the number of amino acids contained in each protein would yield $3 \times 100 \times 2{,}000 = 600{,}000$ nucleotide units in the DNA of an *E. coli*. Depending on the complexity of the organisms, the molecular weights of several millions to several tens of billions have been measured for DNAs (Watson et al. 1987). These molecular weights mean that DNA molecules can contain from several thousand to hundreds of millions of nucleotide units, or enough neighboring base triplets to code for and direct the syntheses of all the different proteins required by an organism.

Let's suppose an *E. coli* bacterium was not able to produce its necessary proteins by such a DNA-directed specific synthesis. Rather, they contained a pool of all 20 amino acids, which randomly reacted and produced proteins with all possible primary structures. If we assume they each contained 100 amino acid residues, then there would be $20^{100} \sim 10^{130}$ different

TABLE 7.3
Approximate Chemical Contents of An Average *E. coli* B/r Cell*

Components	% Total Dry Weight	Amount (g, 10^{-15}) per cell	Molecular Weight	Molecules Per Cell	No. of Different Molecules
Protein	55.0	156	4.0×10^4	2,350,000	1,850
RNA	20.5	58			
23S rRNA		31.0	1.0×10^6	18,700	1
16S rRNA		15.5	5.0×10^5	18,700	1
5S rRNA		1.2	3.9×10^4	18,700	1
t-RNA		8.2	2.5×10^4	198,000	60
m-RNA		2.3	1.0×10^6	1,380	600
DNA	3.1	8.8	2.5×10^9	2.1	1
Lipid	9.1	25.9	705	22,000,000	
Lipopolysaccharide	3.4	9.7	4,070	1,430,000	1
Peptidoglycan	2.5	7.1	$(904)_n$	1	1
Glycogen	2.5	7.1	1.0×10^6	4,300	1
Polyamines	0.4	1.1			
Putrescine		0.83	88	5,600,000	1
Spermidine		0.27	145	1,100,000	
Metabolites, cofactors, ions	3.5	9.9			800+

* Data from Neidhardt, F. C., and Umbarger, H. E., in *Escherichia Coli and Salmonella: Cellular and Molecular Biology*, ASM Press, Washington, DC, 13, 1996.

proteins produced. However, *E. coli* needs only ~2,000 different specific proteins or $(2{,}000/10^{130}) = 2 \times 10^{-125}\%$ of all possible proteins. The rest of the randomly polymerized proteins need to be discarded or be depolymerized to refresh the amino acid pool. What are the consequences of such an inefficient means of protein syntheses?

At least one consequence has to do with the sheer size requirements of such hypothetical organisms. *E. coli*, for example, would have to be ~10^{127} times larger to obtain its necessary proteins by the undirected random Step-Growth synthesis method described above, followed by protein selection and recycling of the remaining unnecessary proteins *via* depolymerization to amino acids. This means that the *E. coli* bacterium that is normally ~20,000 Å long and ~8,000 Å thick (see Figure 7.15) would have to "balloon" to 2.1×10^{36} m in length and 0.86×10^{36} m in width. Stellar distances are measured in light years, i.e., the distance light travels in 1 year or ~10^{16} m. Even moving at the speed of light, a protein in *E. coli* would require ~10^{20} years to navigate from one end of such a galactic *E. coli* to the other. For perspective, compare this to the estimated age of our universe, 10–15 billion years.

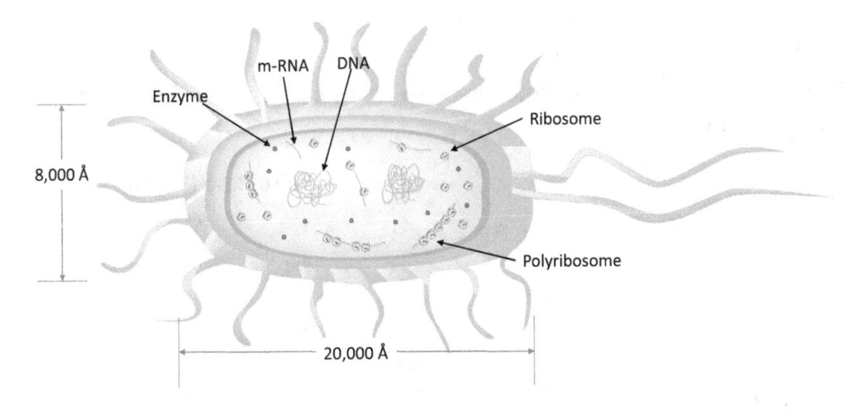

FIGURE 7.15 A schematic drawing of an *E. coli* bacterium with approximate dimensions. (Watson, J. D. et al., *Molecular Biology of the Gene*, 4th Ed., Benjamin-Cummings, Menlo Park, CA, 1987.)

Another consequence faced by *E. coli* and other living organisms relying on the untemplated, undirected random Step-Growth polymerization, aside from their required galactic sizes, would be their astronomical weights. According to Table 7.3, *E. coli* bacteria contain an average of ~1,300 each of their ~2,000 different proteins. Assuming all are 400 amino acid residue proteins, their random synthesis would lead to $20^{400} = 10^{520}$ proteins with different primary sequences. If the amino acids in each of these proteins is assumed to have the molecular weight of an average amino acid residue, which is 93 g/mole (see Table 7.2), then each would weigh $400 \times 93 = 37{,}200$ or $\sim 4 \times 10^4$ g/mole. Thus, 10^{520} structurally different proteins randomly synthesized would weigh $4 \times 10^{524}/6 \times 10^{23} = 6.6 \times 10^{500}$ grams or 6.6×10^{494} tons. The *E. coli*, however, only needs 156×10^{-15} grams of necessary proteins (see Table 7.3). That is to say, the randomly synthesized 10^{520} structurally distinct proteins would weigh 4.2×10^{513} times more than those actually required by a real microscopic *E. coli*.

Needless to say a highly specific, directed, and efficient method for protein syntheses must be used by living organisms, and that method is DNA-directed. Nirenberg and Mathaei (1961), Ochoa (1962), and Khorana (Jones et al. 1966) in a series of ingenious experiments synthesized m-RNAs having particular repeating base triplets (m-RNA reads the DNA message, carries it to the site of protein synthesis on the ribosome, and serves as the template directing protein synthesis). Afterwards, they observed the primary structures of the resulting proteins and established the three base m-RNA genetic code presented in Table 7.4. Based on the genetic code, the 9-base …UUGCCAGGA… nucleotide sequence should and does code, for example, the -Leu-Pro-Gly- protein sequence.

TABLE 7.4
The Genetic Code with Reference to m-RNA Base Triplets Coding for Amino Acids

		Third Base (3′ end)			
First Base (5′ end)	**Second Base**	**U**	**C**	**A**	**G**
U	U	Phe	Phe	Leu	Leu
	C	Ser	Ser	Ser	Ser
	A	Tyr	Tyr	Stop	Stop
	G	Cys	Cys	Stop	Trp
C	U	Leu	Leu	Leu	Leu
	C	Pro	Pro	Pro	Pro
	A	His	His	Gln	Gln
	G	Arg	Arg	Arg	Arg
A	U	Ile	Ile	Ile	Met (Start)
	C	Thr	Thr	Thr	Thr
	A	Asn	Asn	Lys	Lys
	G	Ser	Ser	Arg	Arg
G	U	Val	Val	Val	Val
	C	Ala	Ala	Ala	Ala
	A	Asp	Asp	Glu	Glu
	G	Gly	Gly	Gly	Gly

Note that many of the amino acids are coded by more than a single three-base sequence. This built-in redundancy may offer useful protection against harmful mutations that can alter or transform DNA base sequences. Also notice that neither hydroxyproline nor cystine have codons, so they must be synthesized, respectively, from proline and cysteine. There are also 3-base codons required for beginning and ending the synthesis of each specific protein chain.

Since DNA is a double helix of two complementary strands winding around each other in a right-handed sense, but in opposite directions, one may ask which strand contains the genetic blueprint for synthesizing an organism's proteins? Examination of the genetic code presented in Table 7.4, given with reference to the base sequence triplets of m-RNA, shows that initiation of protein synthesis is caused by an AUG base triplet, while UAA, UAG, and UGA base triplets end or stop a protein's synthesis. Thus, whichever strand of the DNA double helix contains start and stop codons (ATG, TAA, TAG, TGA by replacing U with T) contains the genetic code and is named the coding strand. This is because the TAC, ATT, ATC, ACT complementary triplets that must be present on its antiparallel strand of the DNA double helix do not contain the genetic code to either start or stop protein synthesis and this strand is only used as a template for synthesizing RNA and thus its name "template strand."

An organism's DNA must be able to be reliably reproduced or replicated in order to prevent any alteration in the collection of proteins it codes for, and so that the genetic information of the parent can be transmitted to the offspring. DNA replication is also achieved due to the bases on each nucleotide sugar ring.

Chargraff [as described in Crick (1962)] analyzed the DNAs from a variety of sources and developed the following base composition rules: 1. numbers of A and T bases are always equal, 2. numbers of G and C bases are always equal. As a consequence of numbers 1 and 2, 3. the number of A + G bases must equal the number of T + C bases, and 4. the number of A + T bases and number of G + C bases are not necessarily equal. These observations suggested that there are two distinct purine-pyrimidine pairs, A-T and G-C. Because these DNA base pair relations were observed in all organisms, they provided a strong indication of how DNA can replicate.

X-ray scattering from oriented DNA fibers observed and interpreted by Franklin and Gosling (1953), Wilkins et al. (1953), and Watson and Crick (1953), led to the discovery of DNA's double-helical nature (see Figure 7.16) and how it can replicate itself. Two complementary chains

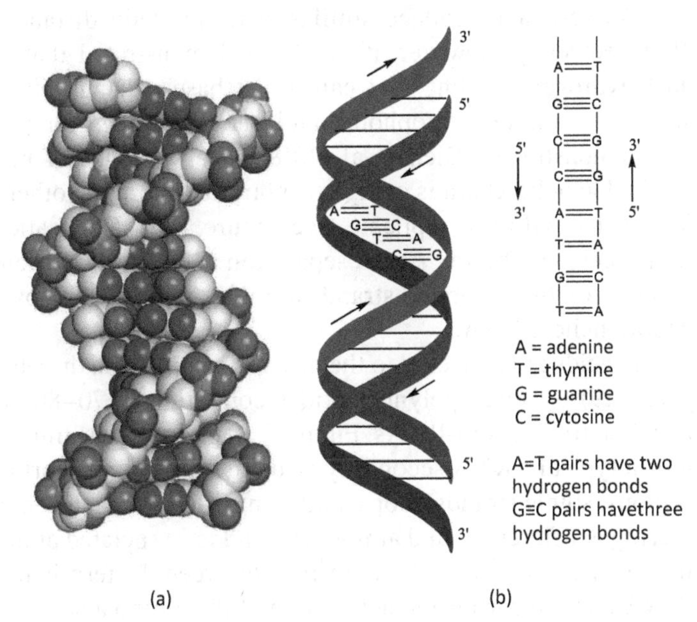

FIGURE 7.16 (a) Space-filling and (b) schematic models of the double helical structure of DNA. Space-filling model generated using PyMol with PDB ID 1BNA, (Structure determined by Drew et al., *Proc. Natl. Acad. Sci. USA* 78, 2179–2183, 1981.) (Schematic models redrawn based on DuPraw, E. J., *Cell and Molecular Biology*, Academic Press, New York, 1968.)

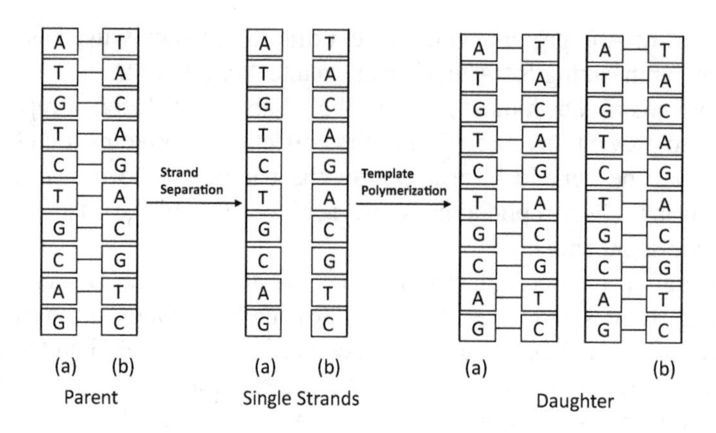

FIGURE 7.17 Schematic of DNA replication. (Redrawn based on Mandelkern, L., *An Introduction to Macromolecules*, Springer-Verlag, New York, 1972.)

with right-handed helical conformations wind around each other in opposite directions to form a DNA double helix, which is shown replicating itself in Figure 7.17.

A-T and G-C base pairings, evident in the DNA double helix in Figures 7.16 and 7.17, are held together by two and three hydrogen bonds, respectively, and produce similar chain-to-chain distances (see Figure 7.18). Recently, however, it has been demonstrated that it may not be just hydrogen-bonding that can direct base pairing, but "shape complementarity and/or hydrophobic and packing forces for pairing" have been demonstrated (Dien et al. 2018). The sequence of bases in one DNA double helix chain is complementary to that in the other chain and is key to its ability to replicate (see Figure 7.17). Replication is a two-stage process combining strand separation and subsequent template polymerization on the separated strands to yield an identical copy of the parent double-helical DNA.

The amino acids are delivered to the m-RNA template on the ribosome (see Figure 7.19) by small polynucleotides consisting of 70–80 nucleotides and called transfer or t-RNAs. Figure 7.20 presents a 2-dimensional diagram of a typical t-RNA secondary structure, including short double helical regions, single strands, loops, and arms. t-RNAs, each containing and delivering a different bound amino acid and its associated anticodon, which are attached by an ester bond formed between the terminal 3′-OH of the t-RNA and the carboxylic acid group of the amino acid.

As illustrated in Figure 7.21, each t-RNA with its bound amino acid is brought to the m-RNA template on the ribosome, and its 3-base anticodon attaches to the complementary m-RNA codon *via* the two or three hydrogen-bonded A-T or G-C base pairs. Although on this diagram the m-RNA base triplets are called codons and the base triplet on the t-RNAs

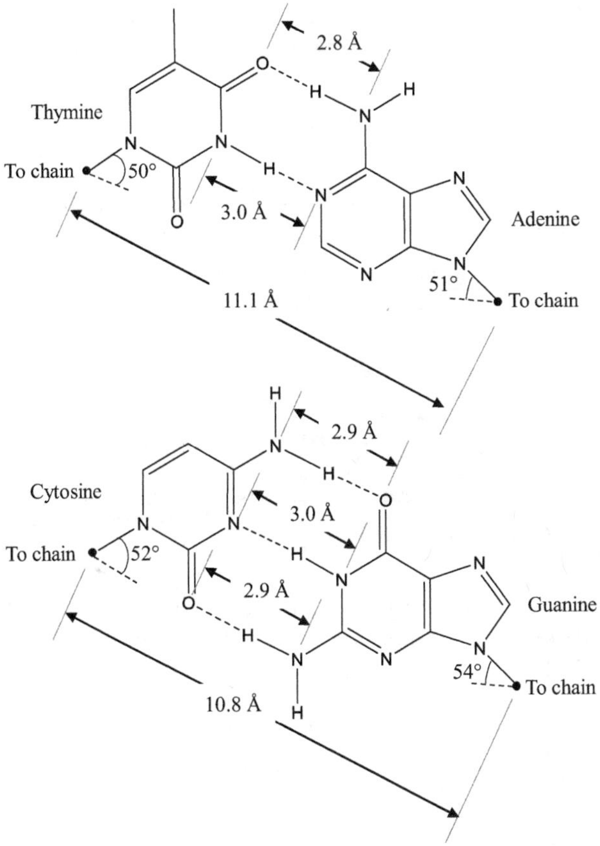

FIGURE 7.18 The pairing of A-T and G-C base pairs in DNA focusing on inter-base hydrogen-bonding and strand-to-strand distances. (Anfinsen, C. *Nobel in Chemistry Lecture*, 1972.)

are called anticodons, remember that the original base triplets on DNA and those on the t-RNAs are the same.

In addition to the amino acids and polynucleotides mentioned here, scores of protein enzymes and other molecules are also involved in the biosynthesis of proteins. However, it is the DNAs and RNAs that are crucially responsible for the specificity, reproducibility, and efficiency of protein syntheses in living organisms. It has been estimated that the biosynthetic assembly of a protein containing 150 amino acid residues can take less than a minute, making it obvious that synthetic chemists have a lot to learn from Nature (Hart et al. 1995).

Very recently it has been proven that the natural base pairs in DNA (A-T and G-C) are not unique in their ability to form and lead to the DNA double helix (Hoshika et al. 2019; Dien et al. 2019). DNA double helices incorporating the synthetic base pairs shown in Figure 7.22 were found to be

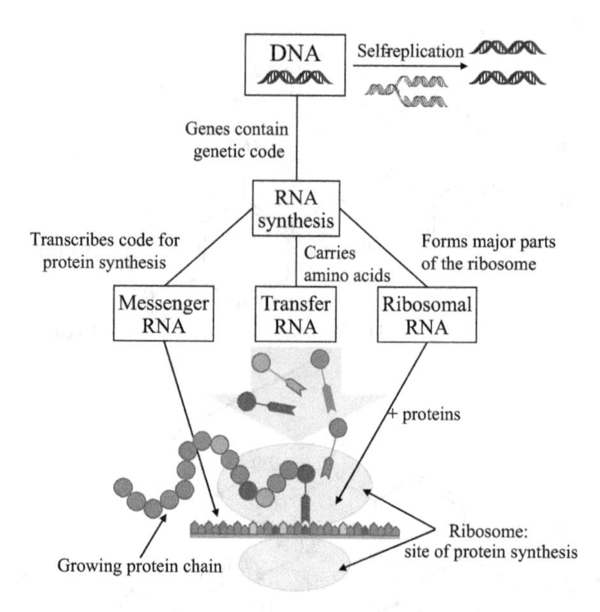

FIGURE 7.19 Schematic diagram of the biological functions of DNA and RNAs. (Mandelkern, L., *An Introduction to Macromolecules*, Springer-Verlag, New York, 1972.)

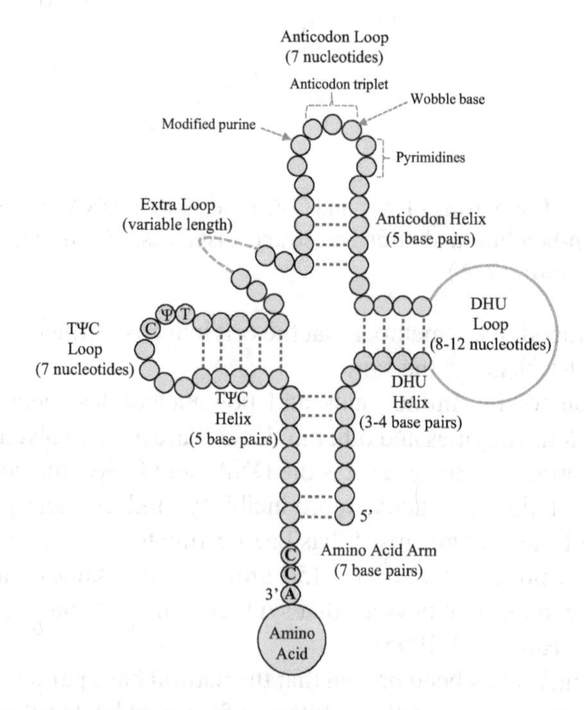

FIGURE 7.20 Generic diagram for t-RNA secondary structure. (Arnot, S., *Prog. Biophys. Mol. Biophys.*, 22, 181, 1971.) The anticodon loop and bound amino acid distinguish the various t-RNAs.

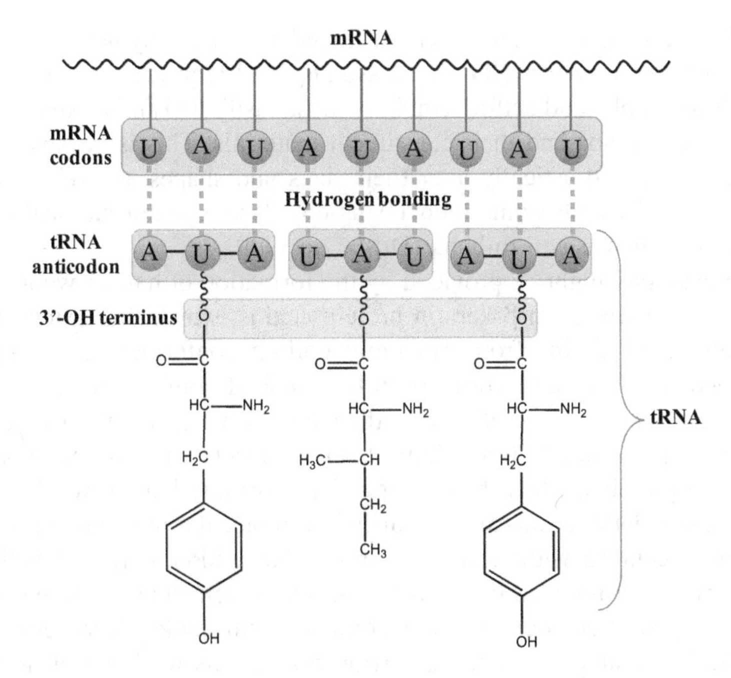

FIGURE 7.21 Schematic depiction of protein synthesis. (Hart, H. et al., *Organic Chemistry: A Short Course*, Houghton Mifflin, Boston, MA, 1995.)

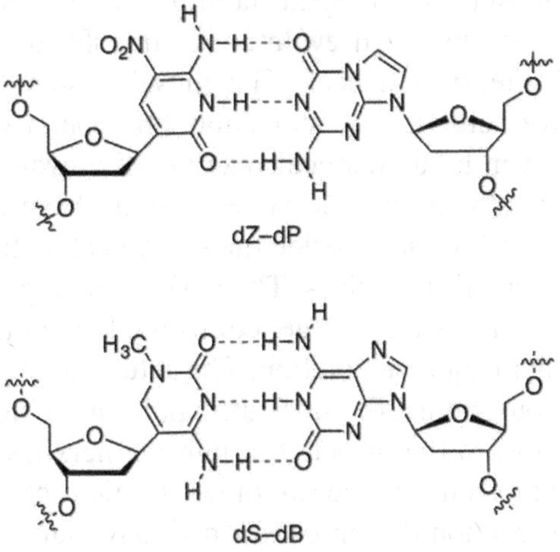

dZ–dP

dS–dB

FIGURE 7.22 dZ-dP and dS-dB unnatural base pairs. (Reprinted with permission from Dien, V. T. et al., *Biochemistry*, 58, 2582–2583, 2019.)

as stable and perform as well as all natural DNA base pairs. This has led to doubt of the long held belief that the natural bases and their pairs were unique and perfect and optimal in their function.

From our discussion of the structures and capabilities of proteins and DNA and RNA polynucleotides, the question arises as to whether life could have originated and been sustained without polymers? It seems likely that the answer is **NO**! The reason is that life processes rely heavily on self-assembly, and self-assembly is most easily and fully expressed in polymeric systems and their constituent materials. This is because polymer chains can drastically alter their sizes and shapes, as well as their interactions, by undergoing unique single molecule conformational phase transitions, such as coil-to-helix (DiMarzio 1999).

An example might be provided by the formation of hair or wool made from α- and silk from β-keratin protein chains, which first extend their as-synthesized chains from randomly coiling conformations to highly extended α-helical and all-*trans* fully extended conformations, respectively. The extended keratins can then intertwine and form two chain α-keratin proto-fibrils (see Figure 7.8) or sheets of hydrogen-bonded β-keratin proteins, which further associate, forming hair or wool or silk fibers. These latter events are primarily intermolecular in nature, but could not have occurred without the intramolecular random-coil to α-helix or all-*trans* conformational transitions of single protein chains. More importantly, they have no counterpart in atomic or small-molecule systems.

Additional support for the assertion that life as we know it here on earth is necessarily macromolecular, is provided by natural muscles. As evidence, we quote a recent review of research on artificial muscle (Otero 2018): "Natural muscles are elegant natural devices developed through millions of years of biological evolution to transform chemical energy into mechanical energy and heat. The muscle contraction is a result of the cooperative actuation of molecular (macromolecular) machines (actin-myosin) driven by a chemical reaction, the hydrolysis of ATP to ADP, and triggered by an electric pulse arriving from the brain (brain order) and liberating Ca_2^+ ions inside the sarcomere: natural muscles are electro-chemo-mechanical devices. The basic actuating unit of a muscular cell (the sarcomere) is a dense gel constituted by molecular chemical machines working in aqueous solution. The actuation of a natural muscle involves: (a) aqueous media, (b) an electric pulse arriving from the brain (the pulse generator) to the muscle through the nervous system, (c) liberation of calcium ions inside the sarcomere (ionic exchange), (d) chemical reactions, (e) reaction-driven conformational changes of the natural polymeric chains (actin and myosin) and d) water exchange for osmotic balance."

To conclude this final chapter, which has briefly described the structures, conformations, and functions of the polymers found in Nature (biopolymers) and compared them to those of synthetic man-made polymers, we offer this brief summary. Synthetic and biological macromolecules share long-chain inherently flexible natures by which they can dramatically alter their sizes and shapes. However, it is the extreme complexity and specificity of biopolymer structures that produces a nearly infinite variety of biopolymer sizes, shapes, properties, and behaviors necessary for all living organisms, and which cannot be achieved with synthetic polymers with their comparatively much simpler microstructures and overall molecular architectures. However, that is not to say that if we begin to form more structurally complicated synthetic polymers, with specific arrangements of their chemically distinct repeat units, that someday we will not begin to mimic some of the very broad range and sophisticated properties and behaviors seen in Nature's biopolymers.

This is likely to begin by constructing them from a limited number of different monomers with preferences for extended conformations and containing chemical groups that would enable interactions, such as hydrogen-bonding, between their chains. This might eventually lead to the development of synthetic material constructs that emulate those of structural proteins like the keratins, collagens, etc.

Again, we see that the same ***Inside*** polymer chain microstructure ←→***Outside*** polymer material property or function approach was successfully used by Nature to form biopolymers that create living organisms. We have attempted to demonstrate here that this approach is also applicable to improving materials formed from synthetic polymers.

A final comment is offered regarding recent investigations that construct "giant molecules" obtained from pre-formed nanoparticle building blocks that are bonded together *via* "click" chemistry (Yin et al. 2017). This approach is inspired by Nature's use of a variety of amino acids to form proteins with exact specific sequences (primary structures) that often fold into a few ordered local secondary structures (conformations), that collectively lead to higher order protein structures (tertiary and quaternary), and determine their biological functions. In "giant molecules," the nanoparticle building blocks replace protein secondary and tertiary structures. They are designed to be rigid and cannot be penetrated by each other, and only surface interactions can influence the overall size and shape of these "giant molecules." Globular proteins, fullerenes, polyhedral oligomeric silsesquioxanes molecules, and polyoxometalates are among the nanoparticles suggested. These may be surface-functionalized and linked together to permit development of giant synthetic mimics of the quaternary packing of globular protein subunits.

We agree with Yin et al. (2017) that "giant molecules provide a general and versatile platform for engineering nanostructures in the field of macromolecular science." However, to hope that designing synthetic polymer constructs in this modular manner will closely imitate Nature's biopolymers seems optimistic at best. For example, we need only remember that the abilities to be repeatedly reproduced and only rarely change and evolve are only observed in biological polymers.

REFERENCES

Alba, K., Bingham, R. J., Gunning, P. O. A., Wilde, P. J., Kontgiorgos, V. (2018), *J. Phys. Chem. Part B*, doi:10.1021acs.jpcb.8b04790.

Anfinsen, C. (1960), *The Molecular Basis of Evolution*, John Wiley & Sons, New York.

Anfinsen, C. (1972, December 11), Studies on the Principles that Govern the Folding of Protein Chains. In *Nobel Lectures, Chemistry 1971–1980*, Frängsmyr, T., Forsén, S., Eds., World Scientific Publishing Co., Singapore, 1993. https://www.nobelprize.org/prizes/chemistry/1972/anfinsen/lecture/.

Arnot, S. (1971), *Prog. Biophys. Mol. Biophys.*, 22, 181.

Babu, M. M. (2016), *Biochem. Soc. Trans.*, 44, 1185–1200.

Babu, M. M., van der Lee, R., de Groot, N. S., Gsponer, J., (2011), *Curr. Opin. Struct. Biol.*, 21, 432–440.

Bartell, L. S., Kohl, D. A. (1963), *J. Chem. Phys.*, 39, 3097.

Baul, U., Chakraborty, D., Mugnai, M. I., Straub, J. E., Thirumalai, D. (2019), *J. Phys. Chem. B*, 123, 3462–3474.

Bell, J. E., Bell, E., (1988), *Proteins and Enzymes*, Prentice-Hall, Englewood Cliffs, NJ.

Bella, J., Brodsky, B., Berman, H. M. (1995), *Structure*, 3, 893–906.

Bonham, R. A., Bartell, L. S. (1959), *J. Am. Chem. Soc.*, 81, 3491.

Brant, D. A. (1997), *Pure Appl. Chem.*, 69, 1885.

Brant, D. A., Christ, M. D. (1990), *Computer Modeling of Carbohydrates*, Amer. Chem. Soc., ACS Symposium Series # 430, Washington, DC, Chapter 4.

Brant, D. A., Flory, P. J. (1965), *J. Am. Chem. Soc.*, 87, 663, 2788, 2791.

Brant, D. A., Miller, W. G., Flory, P. J. (1967), *J. Mol. Biol.*, 23, 47.

Chaney, J. L., Steele, A., Carmichael, R., Rodriguez, A., Specht, A. T., Ngo, K., Li, J., Emrich, S., Clark, P. L. (2017), *PLoS Comput Biol.*, 13, 5, e1005531.

Chatani, E., Hayashi, R., Moriyama, H., Ueki, T. (2002), *Protein Sci.*, 11, 72–81.

Cone, R. D. (2005), *Nat. Neurosci.*, 8, 571.

Cone, R. D. (2006), *Endocr. Rev*,. 27, 736.

Cowan, P. M., McGavin, S. (1955), *Nature*, 176, 501.

Crick, F. H. C. (1962), *Sci. Am.*, 207, 4, 66–77. www.jstor.org/stable/24936717.

Daniel, J. R. (1990), *Concise Encyclopedia of Polymer Science and Engineering*, Kroschwitz, J. L., Ed., John Wiley & Sons, New York, p. 124.

Dickerson, R. E., Geis, I. (1969), *The Structure and Action of Proteins*, Harper and Row, New York.

Dien, V. T., Holcomb, M., Feldman, A. W., Fischer, E. C., Dwyer, T. J., Romesberg, F. E. (2018), *J. Am. Chem. Soc.*, 140, 16115.

Dien, V. T., Holcomb, M., Romesberg, F. E. (2019), *Biochemistry*, 58, 2582–2583.

DiMarzio, E. A. (1999), *Prog. Polym. Sci.*, 24, 329.

Drew, H. R., Wing, R. M., Takano, T., Broka, C., Tanaka, S., Itakura, K., Dickerson, R. E. *Proc. Natl. Acad. Sci. USA*, 78, 2179–2183, 1981.

DuPraw, E. J. (1968), *Cell and Molecular Biology*, Academic Press, New York.

Emrich, Clark, P. L. (2017), *Comput. Biol.*, doi:10.1371/journal.pcbi. 1005531.

Everts, S. (2017), *Chem. & Eng. News*, July, 31, 32.

Flory, P. J. (1969), *Statistical Mechanics of Chain Molecules*, Wiley-Interscience, New York.

Frank, J. (1999), *Sci. Am.*, 86, 428.
Franklin, R. E., Gosling, R. G. (1953), *Nature*, 171, 740, 4356; 172, 156.
Fraser, R. D. B., MacRae, T. P., Rogers, G. E. (1972), Keratins: Their Composition, Structure and Biosynthesis. Charles C. Thomas Springfield, Illinois.
Grassberger, P., Nadler, W., Barkema, G. T. (1998), Eds., *Monte Carlo Approaches to Biopolymers and Proteins*, WorldScience Press, River Edge, NJ.
Gsponer, J., Babu, M. M. (2012), *Cell Rep.*, 2, 1425–1437.
Gurarslan, R., Tonelli, A. E. (2017), *Prog. Polym. Sci.*, 65, 42.
Hart, H., Hart, D. J., Cane, L. E. (1995), *Organic Chemistry: A Short Course*, Houghton Mifflin, Boston, MA.
Herschback, D. R. (1963), International Symposium on Molecular Structure and Spectroscopy, Tokyo, 1962, Butterworths, London, UK.
Hirano, S., Hayashi, M., Nishida, T., Yamamoto, T. (1989), *Chitan and Chitosan*, Skjak-Braek, G., Anthonsen, T., Sanford, P., Eds., Elsevier, London, UK, 743.
Hoshika, S., Leal, N. A., Kim, M. J., Kim, M. S., Karalkar, N. B., Kim, H. J. et al. (2019), *Science*, 363, 884–887.
Iakoucheva, L. M., Brown, C. J., Lawson, J. D., Obradović, Z., Dunker, A. K. (2002), *J. Mol. Biol.*, 323, 573–584.
Jackson, C., O'Brien, J. P. (1995), *Macromolecules*, 28, 5975.
Jones, D. S., Nishimura, S., Khorana, H. (1966), *J. Mol. Biol.*, 16, 454.
Kuchitsu, K. (1959), *J. Chem. Soc. Jpn.*, 32, 748.
Kumaki, J., Kawauchi, T., Okoshi, K., Kusanagi, H., Yashima, E. (2007), *Angew. Chem. Int. Ed.*, 46, 5348.
Li, Y., Douglas, E. P. (2013), *Colloids and Surfaces B.*: Biointerfaces, 112, 42–50.
Lodish, H., Darnell, J. E., Berk, A., Kaiser, C. A., Krieger, M., Scott, M. P., Bretscher, A., Ploegh, H., Matsudaira, P. (2008), *Molecular Cell Biology*, Macmillan.
Mandelkern, L. (1972), *An Introduction to Macromolecules*, Springer-Verlag, New York.
Marqusee, S. Quoting Patricia Clark in Everts, S. (2017), *Chem. Eng., News*, July, 31, 32.
Marsh, R. E., Corey, R. B., Pauling, L. (1955), *Biochem. Biophys. Acta*, 16, 13.
Matheja, J., Degens, E. T. (1971), *Structural Molecular Biology of Phosphate*, Fischer, Stuttgart, Germany.
Mathur, A. B., Tonelli, A. E., Rathke, T. D., Hudson, S. M. (1997), *Biopolymers*, 42, 64.
Mizushima, S. (1954), Structure of Molecules and Internal Rotation, Academic Press, New York.
Natta, G., Corradini, P. (1960), *Nuova Cimento. Suppl.*, 15, 68–82.
Neidhardt, F. C., Umbarger, H. E. (1996), Chapter 3. Chemical Composition of *Escherichia coli*, in *Escherichia coli and Salmonella: Cellular and Molecular Biology*, ASM Press, Washington, DC, 13.
Nirenberg, M. W., Mathaei, J. H. (1961), *Proc. Natl. Acad. Sci. USA*, 47, 1580, 1588.
Ochoa, S. (1962), *Proc. Natl. Acad. Sci. USA*, 47, 1936.
Okano, T., Sarko, A. (1984), *J. Appl. Polym. Sci.*, 29, 4175.
Otero, T. F. (2018), *Int. J. Smart Nano Mater.*, doi:10.1080/19475411.2018.1434246.

Ovchinnikov, S., Park, H., Varghese, N., Huang, P.-S., Pavlopoulos, G. A., Kim, D. E., Kamisetty, H., Kyrpides, N. C., Baker, D. (2017), *Science*, 355, 294.

Pauling, L. (1960), *The Nature of the Chemical Bond*, 3rd Ed., Cornell University Press, Ithaca, New York.

Rathke, T. D., Hudson, S. (1994), *J. Macromol. Sci.*, C34, 375.

Rorrer, G., Hsien, T., Way, J. D. (1993), *Indust. Eng. Chem. Res.*, 32, 2170–2178.

Rose, A. S., Bradley, A. R., Valasatava, Y., Duarte, J. M., Prlić, A., Rose, P. W. (2018), *Bioinformatics*, 34, 21, 3755–3758.

Sarko, A. (1976), *Appl. Polym. Symp.*, 28, 729.

Sasisekharan, V. (1959), *Acta Crystallogr.*, 12, 897.

Sela, M., White, F. H., Anfinsen, C. B. (1957), *Science*, 125, 691.

Shoichet, B. K., Baase, W. A., Kuroki, R., Matthews, B. W. (1995), *Proc. Natl. Acad. Sci. USA*, 92, 452–456.

Slater, G. J., Price, S. A., Santini, F., Alfaro, M. E. (2010), *Proc. R. Soc. B: Biol. Sci.*, 277, 3097.

Tame, J.R., Vallone, B. (2000), *Acta Crystallogr.*, Sect.D 56, 805–811.

Tokura, S., Hasegawa, O., Nishimura, S., Nishi, N., Takatori, T. (1987), *Anal. Biochem.*, 161, 117.

Tompa, P. (2012), *Trends Biochem. Sci.*, 37, 509–516.

Tonelli, A. E. (1970), *J. Chem. Phys.,* 52, 4749.

Tonelli, A. E. (1974), *Analytical Chemistry*, Vol. 3, R. S. Porter and J. F. Johnson, Eds., Plenum, New York, 89.

Tonelli, A. E. (1986), in *Cyclic Polymers*, Semlyen, J. A., Ed., Elsevier, London, UK, Chapter 8.

Tonelli, A. E. (1989), NMR Spectroscopy and Polymer Microstructure: The Conformational Connection, VCH, New York.

Tsuji, H. (2016), *Adv. Drug Deliv. Rev.*, 107, 97.

Uversky, V. N., Oldfield, C. J., Dunker, A. K. (2008), *Annu. Rev. Biophys.*, 37, 215–246.

Voet, D., Voet, J. G., Pratt, C. W. (2008), *Fundamentals of Biochemistry: Life at The Molecular Level.* John Wiley & Sons, Hoboken, NJ.

Wang, B., Yang, W., McKittrick, J., Meyers, M. A. (2016), *Prog. Mater. Sci.*, 76, 229–318.

Watson, H. C. (1969), *Prog. Stereochem.*, 4, 299.

Watson, J. D., Crick, F. H. C. (1953), *Nature*, 171, 737, 964, 4356, 4361.

Watson, J. D., Hopkins, N. H., Roberts, J. W., Weiner, A. (1987), *Molecular Biology of the Gene*, 4th Ed., Benjamin-Cummings, Menlo Park, CA.

Wilkins, M. H. F., Seeds, W. E., Stokes, A. R., Wilson, H. R. (1953), *Nature*, 172, 759, 4382.

Wilson, D. N, Clark, P. L. (2016), *Nature Struct. Mol. Biol.*, 23, 949.

Wilson, E. B., Jr. (1959), *Ad. Chem. Phys.*, 2, 637.

Wright, P. E., Dyson, H. J. (2015), *Nat. Rev. Mol. Cell Biol.*, 16, 18–29.

Yin, G.-Z., Zhang, W.-B., Cheng, S. Z. D. (2017), *Sci. China Chem.*, 60, 338.

Zhao, L., Zhou, X., Rokas, A., Cone, R. D. (2017), *Sci. Repts.*, doi:10.1038/s41598-017-05962-1.

DISCUSSION QUESTIONS

1. Why are the physical properties of amylose starch and cellulose so different despite the fact that both polysaccharides are chains of 1,4-connected D-glucose units?
2. Proteins formed by linking together 20 different amino acids may be considered structurally complex nylon-2s, with just a single carbon separating their amide bonds. They may adopt extended conformations and form fibrous structures or fold up into compact globular conformations. Contrast the behaviors of native and denatured proteins to those of semi-crystalline and molten synthetic nylons.
3. Also compare the inherent conformational flexibilities of protein and nylon chains.
4. What structural information do we know about proteins that are currently unavailable for synthetic polymers, and why is this the case?
5. Proteins are produced by DNA-directed syntheses of only the exact proteins required by each living organism. Why is a directed and highly specific mode of synthesis necessary to produce the necessary set of proteins for an organism?
6. Describe the two critical features of the double-helical structure of DNA.
7. Why is it unlikely that living organisms, such as those on earth, could have been developed or evolved or exist without biopolymers?

Index

Note: Page numbers in italic and bold refer to figures and tables, respectively.